Integrated Electronics: Analog and Digital Circuits

Integrated Electronics: Analog and Digital Circuits

Xavier Myers

Clanrye International
www.clanryeinternational.com

Clanrye International,
750 Third Avenue, 9th Floor,
New York, NY 10017, USA

Copyright © 2023 Clanrye International

This book contains information obtained from authentic and highly regarded sources. All chapters are published with permission under the Creative Commons Attribution Share Alike License or equivalent. A wide variety of references are listed. Permissions and sources are indicated; for detailed attributions, please refer to the permissions page. Reasonable efforts have been made to publish reliable data and information, but the authors, editors and publisher cannot assume any responsibility for the validity of all materials or the consequences of their use.

Trademark Notice: Registered trademark of products or corporate names are used only for explanation and identification without intent to infringe.

ISBN: 978-1-64726-649-3

Cataloging-in-Publication Data

Integrated electronics : analog and digital circuits / Xavier Myers.
p. cm.
Includes bibliographical references and index.
ISBN 978-1-64726-649-3
1. Integrated circuits. 2. Digital electronics. 3. Electronic analog computers--Circuits.
I. Myers, Xavier.

TK7874 .I58 2023
621.381 5--dc23

For information on all Clanrye International publications
visit our website at www.clanryeinternational.com

CLANRYE
INTERNATIONAL

Contents

Preface

An analog circuit refers to a type of electronic circuit that can process any analog data or signal, and generates an output in analog form. These circuits are made up of various devices including capacitors, operational amplifiers, resistors and inductors. Analog circuits work with signals that can range from zero to the maximum power supply voltage. Such circuits have the ability to isolate, attenuate, change, amplify, distort or even convert the original signal to a digital signal. An electronic circuit that processes digital signals is known as a digital circuit. The basic components used in digital circuits include diodes, gates, flip flops and transistors. Both analog and digital circuits are used to process and transmit information such as sound and light from their surroundings in order to create continuous changeable signals. This book elucidates the concepts and innovative models around prospective developments with respect to analog and digital circuits. Researchers and students interested in the study of integrated electronics will be greatly assisted by it.

All of the data presented henceforth, was collaborated in the wake of recent advancements in the field. The aim of this book is to present the diversified developments from across the globe in a comprehensible manner. The opinions expressed in each chapter belong solely to the contributing authors. Their interpretations of the topics are the integral part of this book, which I have carefully compiled for a better understanding of the readers.

At the end, I would like to thank all those who dedicated their time and efforts for the successful completion of this book. I also wish to convey my gratitude towards my friends and family who supported me at every step.

Xavier Myers

1

Operational Amplifier

1.1 Ideal and Practical Op-Amp

An op-amp is a high quality amplifier. It consists of four stages, which are connected in cascaded manner.

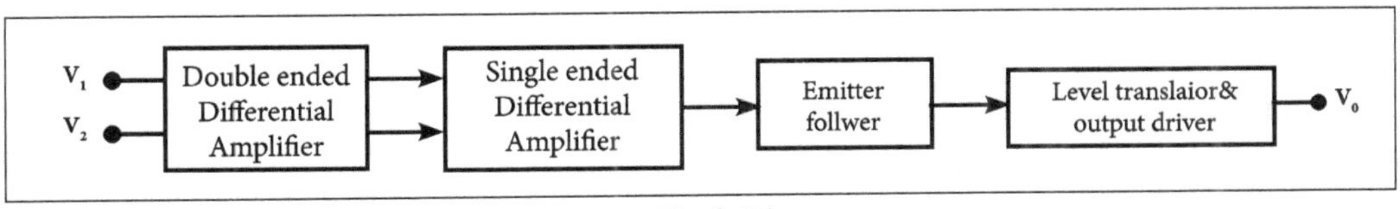

Op-amp Block Diagram.

The first stage of an op. amp is a double ended differential amplifier. This stage provides the maximum voltage gain as it should employ a current source at the common emitter node for good common mode rejection.

This second stage is an intermediate gain stage known as a single ended differential amplifier. It does not need a current source in the emitter. Normally the second stage is needed only to provide some additional gain. Its input resistance must be relatively high to prevent the excessive loading of the first stage.

The third stage is emitter follower that produces unity gain. It consists of high input resistance and also a low output resistance. It matches the output of amplifier stage and input of output stage.

The fourth stage is level translator and the output driver. This stage is used in order to prevent any undesired dc current in the load and increasing the permissible output voltage swing. Hence this supplies large output voltage or current.

An ideal op-amp is expected to have following characteristics:

- Infinite voltage gain.
- Infinite input resistance.
- Zero output resistance.
- Infinite bandwidth.
- Zero noise.

- Zero power supply rejection ratio.
- Infinite common mode rejection ratio.

Practical Op-Amp

None of the above given parameters can be practically realized. A real or practical op-amp has some unavoidable imperfections and hence its characteristics differ from the ideal one. A real op-amp will have non-zero and non-infinite parameters.

Input/Output Characteristics of Practical Op-Amp

The maximum output of an op-amp is limited by dc power supply voltages supplied to op-amp. At most output voltage excursions are limited within $+V_{cc}$ and $-V_{cc}$ where V_{cc} is the power supply voltage. Similarly, output currents from the op-amp are also limited by these saturation voltages.

Below is an example of op-amp with gain 1, 28, 571 and power supply voltages equal to 9 volts. When input voltages exceeds + or -70 micro volts op-amp starts to exhibit nonlinear characteristics and output voltage saturates at + or -9 volts.

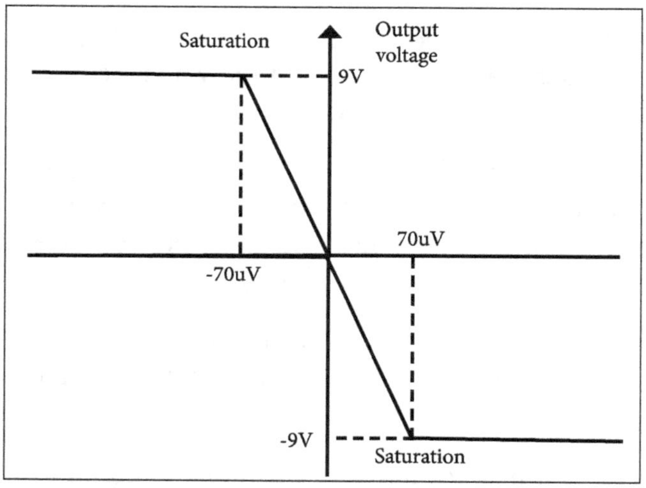

Input/output characteristics of Practical Op-amp.

1.2 Op-Amp, 741 Op-Amp, DC and AC: Characteristics

Op-Amp Characteristics

(a) Supply Voltage

Two types of supply are used for op amps:

- Dual supply voltage.

- Single supply voltage.

Many older type op amps, uses dual supply (+VS and -VS) often in the 12 to 18V range. Single voltage supplies have grown in popularity with the increase in portable devices where dual supplies using multiple batteries are more expensive to implement.

(b) Frequency Response

In the ideal amplifier, its frequency response should be infinite it will amplify all type of frequencies equally. In practical amplifiers, this is difficult to achieve and not always desirable, but op amps have extremely wide and easily variable bandwidths.

(c) Open Loop Voltage Gain

The below graph shows that the open loop gain (with no feedback) at very low frequencies is huge. Open loop gain refers to the maximum AC gain at very low frequencies. The below graph shows that open loop voltage gain of about 126dB, but at frequencies above a few Hz, gain begins to fall rapidly at 20dB/decade.

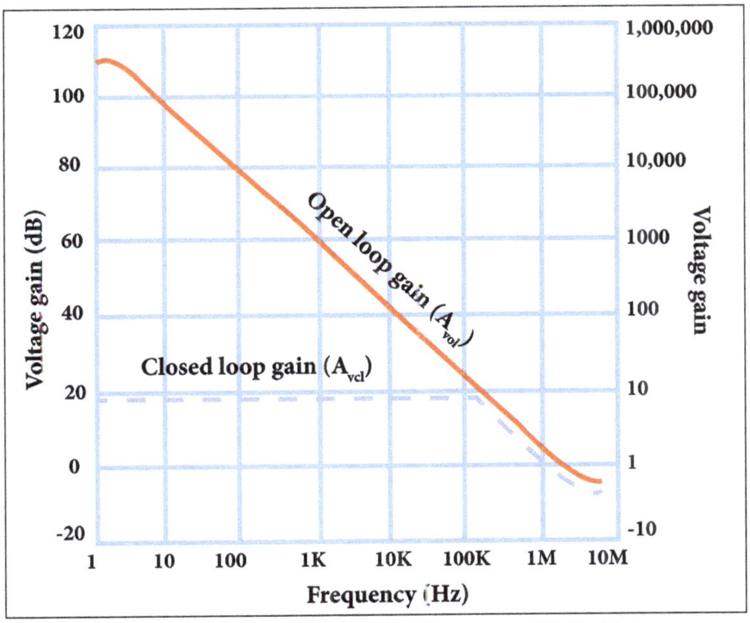

Frequency response of a typical Op Amp (LMC660).

(d) Large Signal Voltage Gain

This is the open loop voltage gain measured at DC with the amplifier producing the large voltage output.

(e) Closed Loop Voltage Gain

Practically, huge gain of an op amp is greatly reduced by applying an appropriate amount of negative feedback. The dotted line in the above graph shows the response of the op amp with negative feedback. The gain has been reduced to 20dB, a closed loop voltage gain (Avcl) of x10, which has produced a flat response from 0Hz to 140kHz.

(f) Gain Bandwidth Product

The term Gain Bandwidth Product is often used to describe the possible combinations of gain and bandwidth. Bandwidth indicated by the Gain Bandwidth Product has been applied to small signals.

For large AC signals with fast rising and falling edges, the bandwidth may be further reduced by the Slew Rate. Then, the Power Bandwidth becomes more relevant.

(g) Maximum Differential Input

This is the maximum voltage that can be applied between the two inputs. In some devices, this can be equal to the supply voltage.

(h) Input Resistance

It is the resistance looking into the input terminals with the amplifier operating without feedback (open loop). For bipolar devices, it ranges from 1MΩ to 10MΩ. For FET and CMOS types, it ranges up to 10^{12} Ω or more.

(i) Input Offset Current

Currents flowing into two inputs must be ideally zero. But for practical op amps, small input currents exists and may also be different. Unequal currents causes different voltages at the inputs.

When this small difference in voltage is amplified, it causes the output to be other than zero. To overcome this effect, an input offset voltage can be applied between the inputs to correct the output voltage to zero. Typical values for bipolar op amps would be ± 1mV ranging up to 15mV for FET types.

(j) Temperature Coefficients

Both the input offset current and input offset voltages are affected by the changes in temperature coefficients. The temperature coefficient of input offset current is measured in nA or pA/°C where as the temperature coefficient of input offset voltage is usually measured in µV/°C.

(k) Slew Rate

The Slew rate describes how fast the output voltage can change in response to an immediate change in voltage at the input. Higher the value (in V/µs) of slew rate, faster the output can change.

If a square wave is applied to the input, the output should also be a square wave but the fast rising and falling edges of the square wave causes the amplifier to oscillate for a short time after the rise or fall edges occurrence. To prevent this, the op amp's internal circuitry contains a small amount of compensation capacitance.

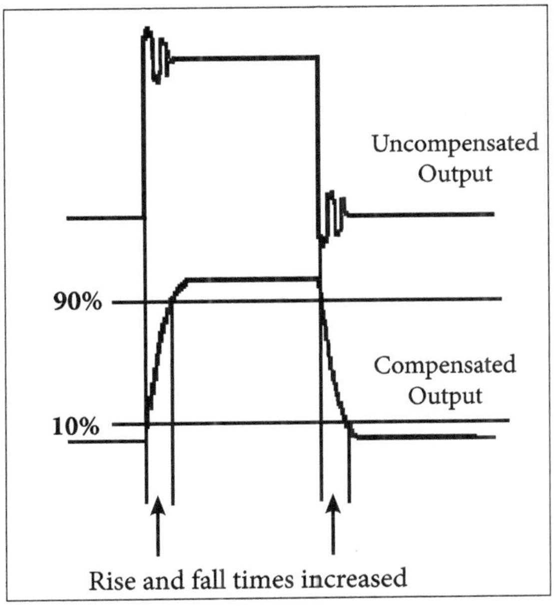

Slew Rate.

It slows down the rate of change by acting as a time constant. This compensation also limits the slew rate of the op amp. If the slew rate of the amplifier cannot keep up with the fastest rate of change of the signal, some distortion will be produced.

For amplifying the large amplitude signals, an op-amp needs to have sufficiently high value of slew rate to cope with the greatest possible rate of voltage change. If the largest possible voltage swing and the highest frequency of the signal are known, the minimum required slew rate for the op amp can be calculated using the below given formula,

$$\text{Slew Rate}\left(V/\mu s\right) = 2\pi f\, V_{pk}$$

Where,

f = The highest signal frequency (Hz).

V_{pk} = The maximum peak voltage of the signal.

(l) Power Bandwidth

In an op amp operating from a ± 15V supply, the power bandwidth would be specified as the frequency range in which a ± 10V swing can be measured at the output with a total harmonic distortion of less than 5%.

1.2.1 DC and AC Characteristics

DC Characteristics

An ideal op amp draws no current from the source and its response is also independent of temperature. But in a real op amp, current is taken from the source to input terminals.

The two input's respond differently to current and voltage due to mismatch in transistors.

The non-ideal characteristics are:

- Input bias current.
- Input offset current.
- Output offset voltage.
- Thermal drift.

Input Bias Current

The Op-amp's input is a differential amplifier, which may be made of BJT or FET. In either case, the input transistor must be biased into their linear region by supplying currents into the bases by the external circuit.

In ideal Op-amp, no current is drawn from input terminals. But practically, op-amp input terminals do conduct a small value of AC current to bias input transistors. Manufacturers specify the input bias current IB as the average value of base currents entering into the terminals of an op-amp as shown in the below figure:

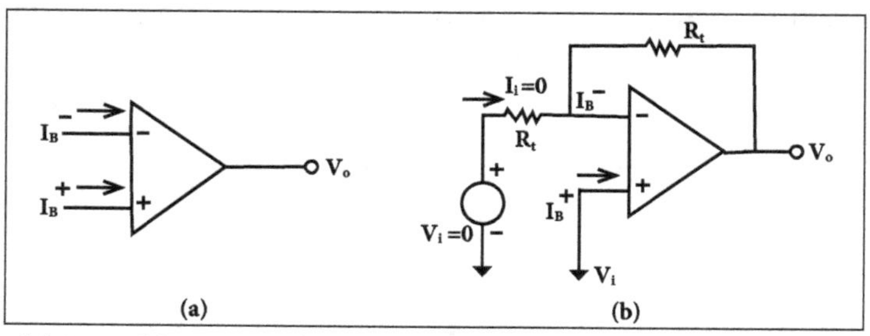

(a) Input bias currents (b) Inverting amplifier with bias currents.

Bias current is the input current. The transistor (internal transistor of the input differential stage) needs to produce an amount of output voltage.

So,

$$I_B = \frac{I_B^+ + I_B^-}{2} \qquad ...(1)$$

For IC 741, a bipolar op-amp, the bias current is 500 nA or less. Consider the basic inverting amplifier. If the input voltage Vi is set to zero volts, VO should be zero volts.

Instead, we find that the output voltage is offset by,

$$V_O = (I_B)R_f \quad ...(2)$$

For example,

If $R_f = 1M\Omega$, $I_B = 500$ nA, then $V_O = 500$ mV with zero input. Output voltage is 500 mV. This is unacceptable. This effect can be compensated as in the below figure by adding a resistor Rcomp between the non-inverting input terminal and ground.

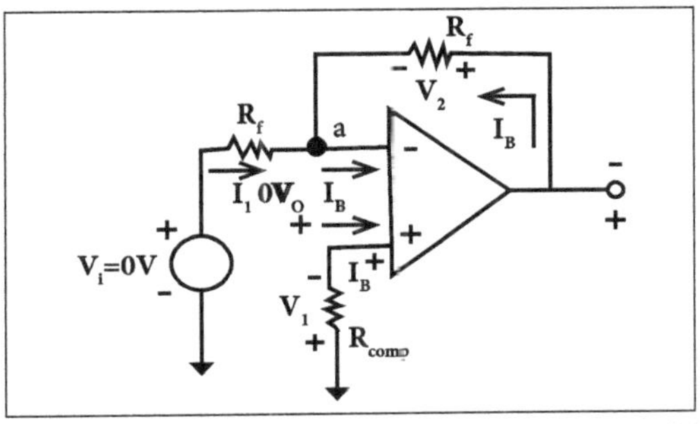

Bias Current Compensation in an Inverting Amplifier.

Current I_B^+ flowing through Rcomp develops a voltage V_1 across it.

Then, by KVL, we get as,

$$-V_1 + 0 + V_2 - V_o = 0$$

$$V_o = V_2 - V_1 \quad ...(3)$$

By selecting proper value of Rcomp, V_2 can be cancelled with V1 and output V_O will be zero.

∴ The value of Rcomp is derived as,

$$V_1 = I_B + R_{comp}$$

$$I_B^+ = \frac{V_1}{R_{comp}} \quad ...(4)$$

The node 'a' is at voltage $(-V_1)$, because the voltage at the non-inverting input terminal is $(-V_1)$. So with $V_i = 0V$, we get as,

$$I_1 = \frac{V_i - (-V_1)}{R_1} = 0 + \frac{V_1}{R_1} = \frac{V_1}{R_1} \quad [\therefore V_i = 0]$$

$$I_1 = \frac{V_1}{R_1} \quad ...(5)$$

Also,

$$I_2 = \frac{V_2}{R_f} \quad ...(6)$$

For compensation, V_O should be zero for $V_i = 0V$, i.e. $V_1 = V_2$

$$\therefore \; I_2 = \frac{V_1}{R_f} \quad ...(7)$$

KCL at node 'a' gives,

$$\Rightarrow I_1 + I_2 = I_B$$

Substituting for I1 and I2,

$$I_B^- = \frac{V_1}{R_1} + \frac{V_1}{R_f}$$

$$\therefore I_B^- = V_1 \left(\frac{R_1 + R_f}{R_1 R_f} \right) \quad ...(8)$$

Assuming $I_B^- = I_B^+$ and using equation (4) and (8) we get,

$$I_B^+ = \frac{V_1}{R_{comp}}$$

$$\therefore \; \frac{V_1 (R_1 + R_f)}{R_1 R_f} = \frac{V_1}{R_{comp}} \quad ...(9)$$

$$R_{comp} = R_1 \| R_f = \frac{R_1 R_f}{R_1 + R_f}$$

To compensate bias currents, R_{comp} should be equal to parallel combination of resistors R_1 and R_f. The effect of input bias current in an amplifier can be compensated by placing the compensating resistor R_{comp}.

Input Offset Current

The difference in magnitudes of I_B^+ and I_B^- is called as input offset current (I_{os}). Bias current compensation will work if both I_B^+ and I_B^- are equal. Since the input transistors are not identical, there is a small difference between I_B^+ and I_B^-. This difference is called as input offset current (I_{os}) and it can be written as,

$$|I_{OS}| = I_B^+ - I_B^- \qquad ...(10)$$

The absolute value of sign indicates that there is no way to predict which of the bias currents will be larger.

Offset current Ios for BJT op-amp is 200 nA. Offset current Ios for FET is 10 pA. Even with bias current compensation, Ios will produce output voltage when Vi = 0V.

$$V_1 = I_B^+ R_{comp} \qquad ...(11)$$

And,

$$I_1 = \frac{V_1}{R_1} = \frac{I_B^+ R_{comp}}{R_1} \qquad ...(12)$$

KCL at node 'a' $\Rightarrow I_1 + I_2 = I_B^-$

$$I_2 = I_B^- - I_1.$$

$$I_2 = I_B^- - \left(\frac{I_B^+ R_{comp}}{R_1}\right) \qquad ...(13)$$

$$V_O = V_2 - V_1$$

$$V_O = I_2 R_f - V_1$$

$$= I_2 R_f - I_B^+ R_{comp}$$

$$= \left(I_B^- - \frac{I_B^+ R_{comp}}{R_1}\right) R_f - I_B^+ R_{comp} \qquad ...(14)$$

substituting for $R_{comp} = R_1 \| R_f$

$$V_O = I_B^- R_f - I_B^+ R_{comp} \left[\frac{R_f}{R_1} + 1\right]$$

$$= I_B^+ R_f - I_B^+ \frac{R_1 R_f}{R_1 + R_f} \left[\frac{R_1 + R_f}{R_1}\right]$$

$$V_O = R_f \cdot \left[I_B^{+} - I_B^{-} \right] \qquad ...(15)$$

$$V_O = R_f \cdot I_O \qquad ...(16)$$

So even with current bias compensation and with feedback resistor of 1MΩ, a 741 BJT Op-amp has an output offset voltage,

$$V_O = 1\text{M}\Omega \times 200\text{ nA} = 200\text{ mV, With zero input voltage}$$

It can be seen that effect of offset current can be minimized by having a small value of Rf. Unfortunately, R1 must be kept large to have high input impedance. With R1 large, Rf also should be kept large so as to have reasonable gain.

This will allow large feedback resistance, while keeping the resistance to ground low as shown in dotted in the below given network.

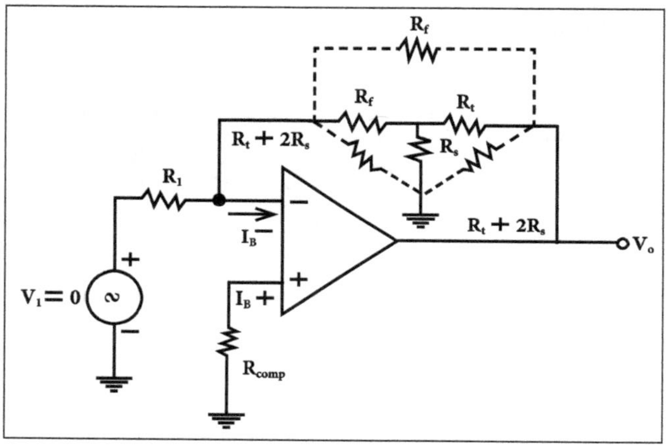

Inverting amplifier with 'T' feedback network.

The 'T' Network provides a feedback signal as if the network were a single feedback resistor. By T to Π conversion,

$$R_f = \frac{R_t^2 + 2R_t R_s}{R_s} \qquad ...(17)$$

$$R_f R_s = R_t^2 + 2R_t R_s$$

$$R_t^2 = R_f R_s - 2R_t R_s$$

$$R_t^2 = R_s (R_f - 2R_t)$$

To design a T-network,

$$R_t << \frac{R_f}{2} \qquad ...(18)$$

Then calculate $R_s = \frac{R_t^2}{R_f - 2R_t}$ ...(19)

Input Offset Voltage

Whenever both the input terminals of the op-amp are grounded ideally, the output voltage should be zero. However the practical op-amp shows a small non-zero output voltage.

To make this output voltage to be zero, we may have to apply a small voltage at the input terminals.

This voltage is called input offset voltage V_{os}, since positive terminal (+) voltage is around V_1=0V.

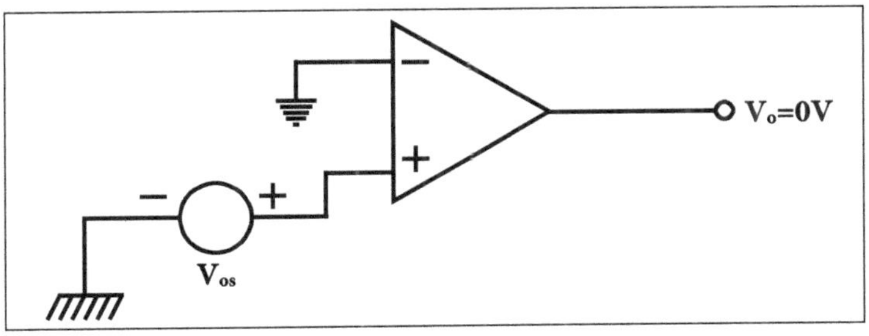

Op-amp Showing Input Offset Voltage.

The voltage V_2 at (-) input terminal is given by,

$$V_2 = \left(\frac{R_1}{R_1 + R_f}\right) V_o \qquad ...(20)$$

$$V_o = \left(\frac{R_1 + R_f}{R_1}\right) V_2$$

$$V_o = \left(1 + \frac{R_f}{R_1}\right) V_2 \qquad ...(21)$$

Since,

$$V_{ios} = |V_1 - V_2| \qquad ...(22)$$

And,

$$V_1 = 0\text{ V}$$

$$V_{OS} = |0 - V_2| = V_2$$

$$\therefore\ V_o = \left(1 + \frac{R_f}{R_1}\right) V_{ios} \qquad ...(23)$$

Thus, the output offset voltage of an op-amp in closed loop configuration (inverting and non-inverting) is given by equation (23).

Total Output Offset Voltage

The total output offset voltage VoT could be either more or less than the offset voltage, produced at the output due to input bias current or input offset voltage alone. This is because the input offset voltage Vios and input bias current IB could be either positive or negative with respect to ground.

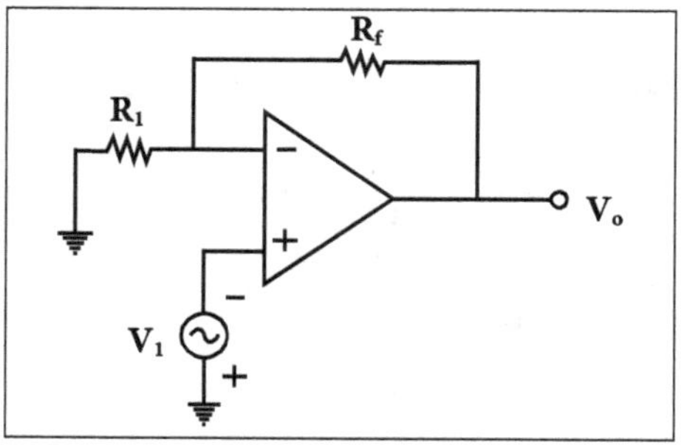

(a) Non Inverting amp.

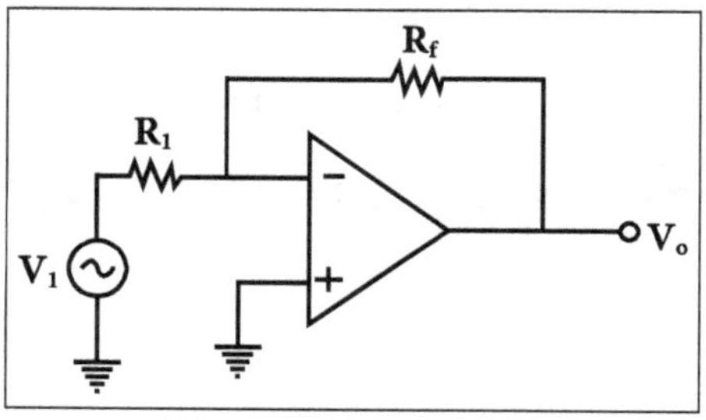

(b) Inverting amplifier.

Without any compensation technique used, the maximum offset voltage at output of inverting and non-inverting amplifier of above figure (a) and (b) is given below,

$$V_{oT} = \left(1 + \frac{R_f}{R_1}\right) V_{ios} + R_f\, I_B \qquad \text{...(24)}$$

With Rcomp in the circuit, the total output offset voltage will be given by,

$$V_{oT} = \left(1 + \frac{R_f}{R_1}\right) V_{OS} + R_f . I_{ios} \qquad \text{...(25)}$$

Many op-amps provide offset compensation pins to nullify the offset voltage. The below figure gives the connections for 741 op-amp.

The manufacturers recommend a 10 k Ω potentiometer be placed across offset null pins. Pin 1 and 5 and the wiper is connected to the negative supply pin 4. The position of wiper is adjusted to nullify the output offset voltage.

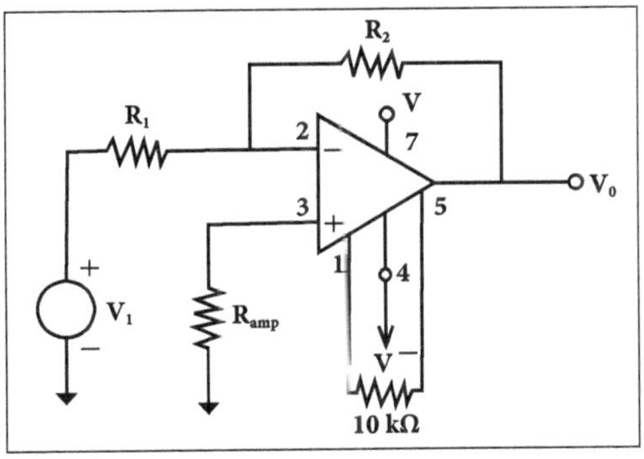

Offset null pin connection for μA741.

AC Characteristics

For small signal sinusoidal (AC) applications, we have to know the AC characteristics such as frequency response and slew rate.

Frequency Response of an OP-Amp

Ideally an open loop op-amp has infinite bandwidth. It means that, if it's AoL = 90dB with DC signal, its gain should remain the same 90dB through audio and onto high frequencies. But with practical op-amp, gain decreases (rolls off) at higher frequencies.

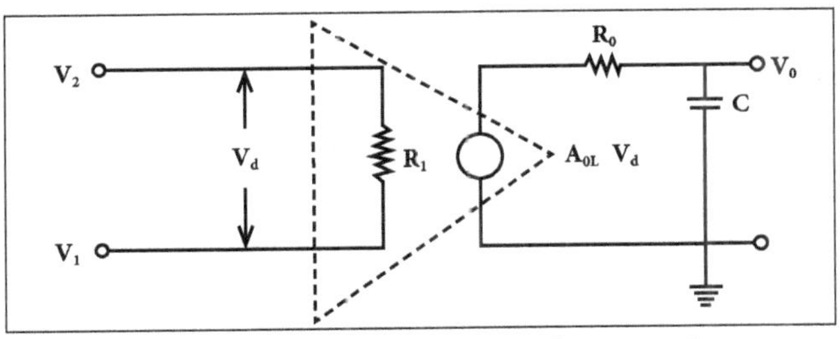

High frequency of an op-amp with single corner frequency.

The reason for the above is that there must be a capacitance component in the equivalent circuit of op-amp. This capacitance is due to the physical characteristics of device (BJT or FET) used and internal construction of op-amp. For the op-amp with one corner frequency all the capacitor effects can be represented by a single capacitor C as in the below figure.

Due to single RoC, there is one pole and obviously one −20dB/ decade rolls off comes into effect. The open loop gain of an op-amp with only one corner frequency is obtained from the above figure as,

$$v_o = \frac{-jXc}{R_o - jXc} A_{OL} . Vd \qquad ...(26)$$

$$A = \frac{V_o}{V_d} = \frac{A_{OL} .1}{1 + j2\pi fR_o C} \qquad ...(27)$$

$$A = \frac{A_{OL}}{1 + j(f/f_1)}$$

or,

$$A = \frac{A_{OL}}{\sqrt{1 + j\left(\frac{f}{f_1}\right)^2}}$$

Where,

$$f_1 = \frac{1}{j2\pi fR_0 C} \quad ...(28)$$

Where,

f_1 - corner frequency of the op-amp.

The magnitude and phase angle of the open loop voltage gain can be written as,

$$|A| = \frac{A_{OL}}{\sqrt{1 + j\left(\frac{f}{f_1}\right)^2}} \quad ...(29)$$

$$\phi = -\tan^{-1}\left(\frac{f}{f_1}\right) \quad ...(30)$$

The magnitude and phase characteristics are shown in the below figure:

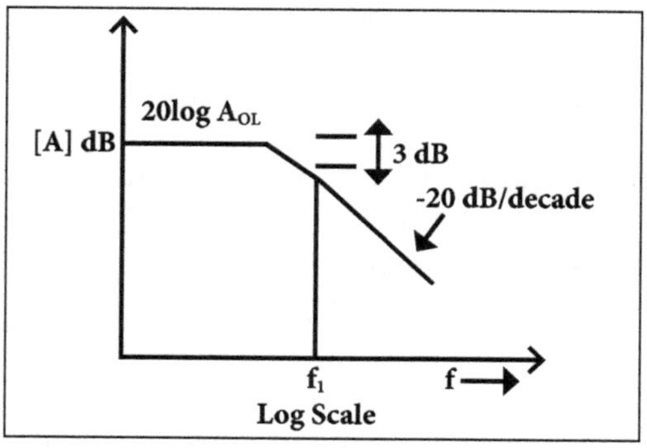

Open Loop Magnitude Characteristics in Semi-Log Paper.

From the above graph,

- For frequency $f < f_1$, the magnitude of the gain is 20 log AOL in dB.
- For $f = f_1$, the gain is 3 dB down from the dc value of AOL in dB. This frequency f1 is called corner frequency.
- For $f >> f_1$, the gain rolls off at the rate of –20 dB/decade (or) 6dB/octave.

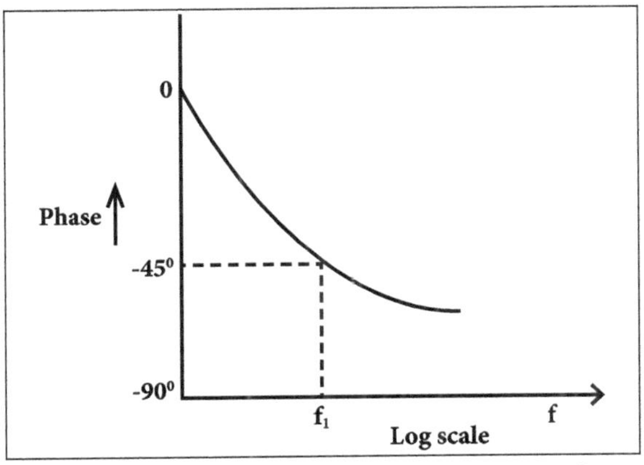

Phase Characteristics For An Op-Amp With Single Break Frequency.

From the phase characteristics, the phase angle is zero at frequency $f = 0$. At f1, phase angle is −45°(lagging) and at infinite frequency phase angle is -90°. This shows that a maximum of 90° phase change can occur in an op-amp with a single capacitor.

The voltage transfer function in s-domain can be written as,

$$A = \frac{A_{OL}}{1 + j\frac{f}{f_1}}$$

$$= \frac{A_{OL}}{1 + j(\omega / \omega_1)} = \frac{A_{OL} \cdot \omega_1}{j\omega + \omega_1}$$

$$= \frac{A_{OL} \cdot \omega_1}{s + \omega_1}$$

Due to the number of RC pole pairs, there will be a number of different break frequencies. The transfer function of an op-amp with 3 different break frequencies can be assumed as,

$$A = \frac{A_{OL}}{\left(1 + j\frac{f}{f_1}\right)\left(1 + j\frac{f}{f_2}\right)\left(1 + j\frac{f}{f_3}\right)} \quad \text{...(31)}$$

$$0 \geq f_1 < f_2 < f_3$$

$$A = \frac{A_{OL} \cdot \omega_1 \omega_2 \omega_3}{(s + \omega_1)(s + \omega_2)(s + \omega_3)} \quad \text{...(32)}$$

With $0 < \omega_1 < \omega_2 < \omega_3$

Approximation of open loop gain V_s frequency curve is shown in the below figure:

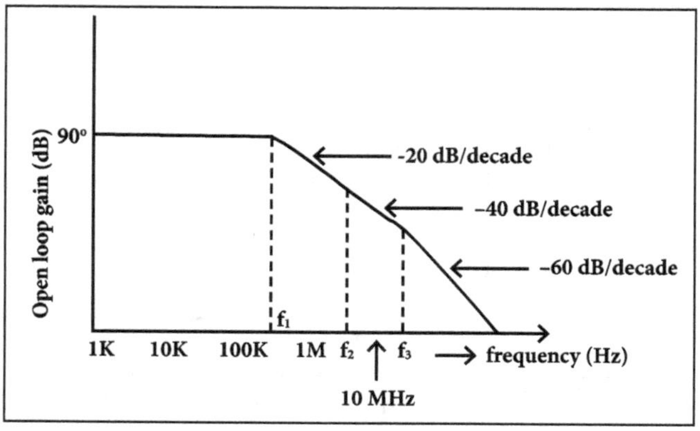

Approximation of Open Loop Gain Vs Frequency Curve.

Approximation of open loop gain Vs frequency curve is shown in the above figure. For frequencies from 0 Hz to 200 KHz, open loop frequency response is constant (90 db).

From 200 kHz - 2MHz, gain drops from 90 dB to –70 dB which is at a -20 dB/decade. From 2 MHz – 20 MHz - the roll off rate is –40 dB /decade.

As frequencies increases, cascading effect of RC pairs (poles) come into effect and roll-off rate increases successively by -20 dB/ decade at each corner frequency.

1.2.2 Characteristics of 741 Op-Amps

The following are the some of the Characteristics of 741 Op-amp:

- Low power consumption.
- Short circuit and overload protection provided.
- No latch-up problem.
- Large common mode rejection ratio (CMRR) and differential voltage ranges.
- No external frequency compensation is required. It also does not need any external compensation for phase component. This simplifies the circuit design and minimizes the number of components used.
- In theory, the dc output voltage will be zero if both the inputs of the 741 IC are connected to the ground. But in practice, a small dc output may appear due to minor internal unbalances. It is usually unnoticed in normal applications. But for critical conditions, the output voltage can be set precisely to zero by connecting a 10K potentiometer between terminals marked offset-null.

1.3 Modes of Operation

Inverting and Non-Inverting Amplifier

In this inverting amplifier circuit, the operational amplifier is connected with the feedback to produce a closed loop operation. When dealing with operational amplifiers, there are two important rules to remember about the inverting amplifiers, they are: no current flows into input terminal and that V_- is always equal to V_2. But, in real world op-amp circuits both of these rules are slightly broken.

Because the junction of the input and feedback signal is at same potential as the positive input, the junction at zero volts or ground, then the junction is called as the virtual earth.

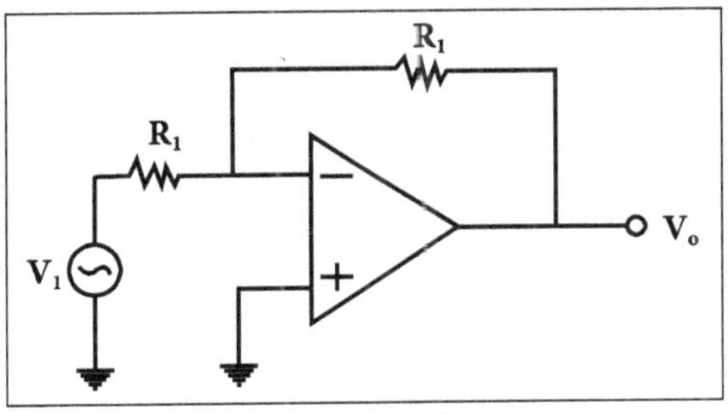

Inverting Amplifier.

Because of the virtual earth node, the input resistance of the amplifier is equal to the value of the input resistor, Rin and the closed loop gain of the inverting amplifier may be set by ratio of two external resistors.

There are two important rules to remember about inverting amplifiers or any operational amplifier for that matter and they are,

- No current flows into the input terminals.
- The Differential input voltage is zero as $V_1=V_2=0$.

Then by using the two rules, we can able to derive the equation for calculating the closed-loop gain of an inverting amplifier using first principles.

Current (i) flows through the resistor network is shown as below,

$$i=\frac{V_{in}-V_{out}}{R_{in}+R_f}$$

Therefore,

$$i=\frac{V_{in}-V_2}{R_{in}}=\frac{V_2-V_{out}}{R_f}$$

$$i=\frac{V_{in}}{R_{in}}-\frac{V_2}{R_{in}}=\frac{V_2}{R_f}-\frac{V_{out}}{R_f}$$

So,

$$\frac{V_{in}}{R_{in}}=V_2\left[\frac{1}{R_{in}}+\frac{1}{R_f}\right]-\frac{V_{out}}{R_f}$$

And as,

$$i=\frac{V_{in}-0}{R_{in}}=\frac{0-V_{out}}{R_f} \qquad \frac{R_f}{R_{in}}=\frac{0-V_{out}}{V_{in}-0}$$

The closed Loop Gain (Av) is given as, $\frac{V_{out}}{V_{in}}=-\frac{R_f}{R_{in}}$

Then, the closed-loop voltage gain of the inverting amplifier can be represented as,

$$\text{Gain}(Av)=\frac{V_{out}}{V_{in}}=-\frac{R_f}{R_{in}}$$

And this can be transposed to give Vout as,

$$V_{out}=-\frac{R_f}{R_{in}}\times V_{in}$$

The negative sign indicates the inversion of the output signal with respect to the input as it is 180° out of phase. This is due to the feedback being negative in value. The equation for the output voltage V_{out} also shows that the circuit is linear in nature for a fixed amplifier gain as $V_{out} = V_{in}$ x Gain. This property is very useful for converting a smaller sensor signal to a much larger voltage.

Another useful application of an inverting amplifier is that of trans-resistance amplifier circuit.

Non Inverting Amplifiers

In non-inverting circuit configuration, input impedance Rin has now increased to infinity and its feedback impedance R_f has reduced to zero.

The output is connected directly back to the negative inverting input so the feedback is 100% and V_{in} is exactly equal to V_{out} giving it a fixed gain of 1 or unity. As the input voltage Vin is applied to the non-inverting input the gain of the amplifier is given as,

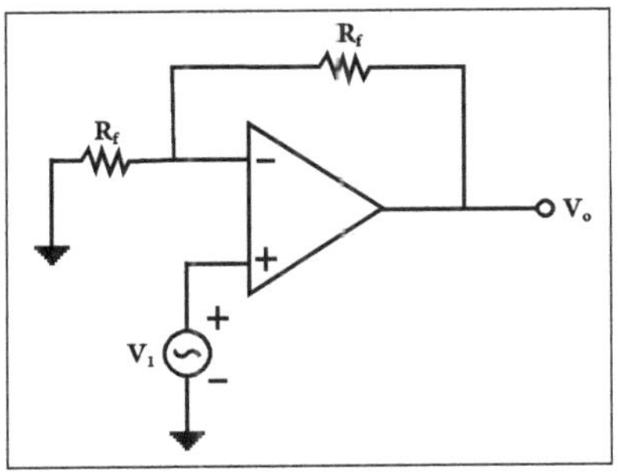

Non Inverting amplifiers.

$$V_{out} = A(V_{in})$$

$$(V_{in} = V_{+}) \quad \text{and} \quad (V_{out} = V_{-})$$

Therefore Gain,

$$(Av) = \frac{V_{out}}{V_{in}} = +1$$

1.3.1 Differential Amplifier

The below figure shows the circuit of differential amplifier. As the name suggests, this op-amp configuration amplifies difference of the two input signals.

If the two input signals are the same, the output must be zero ideally. To quantify the quality of amplifier, the term Common Mode Rejection Ratio (CMRR) is defined. It is the ratio of output voltage corresponding to the difference of two input signals to that of the output voltage corresponding to common part of two signals. A good op-amp has a high CMRR.

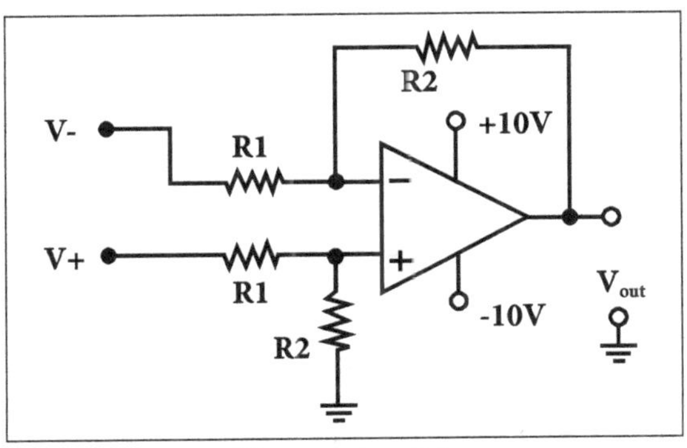

Difference Amplifier.

1.3.2 Instrumentation Amplifier

Physical quantities such as temperature, humidity, light intensity, water flow etc. are measured with the help of transducers. An instrumentation amplifier is used to amplify the output signal produced by a transducer, so that it can drive the indicator or display system.

The below figure shows the simplified form of such a system.

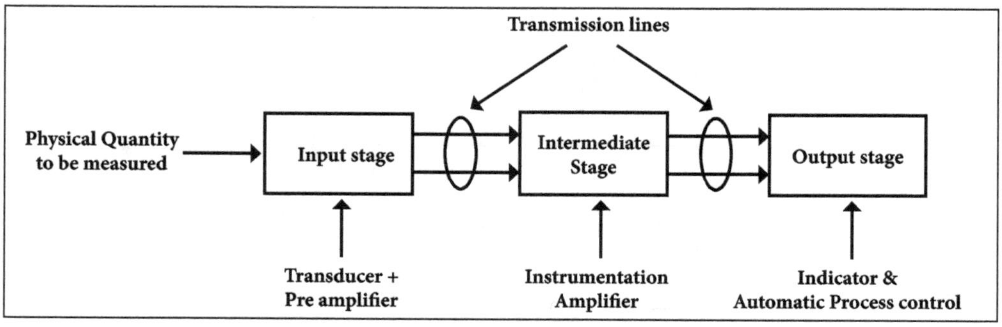

Simplified Form of Instrumentation Amplifier.

The input stage is composed of a preamplifier and some sort of transducer, depending on the physical quantity to be measured. The output stage may use devices such as meter, oscilloscope, charts or magnetic recorders. Some transducer produce outputs with sufficient strength to permit their use directly, many do not.

To amplify the low-level output signal of the transducer, so that it can drive the indicator or display, an instrumentation amplifier is employed.

Thus, instrumentation amplifiers are used in monitoring and controlling of the physical quantities in the industrial process for measurement and control of temperature, humidity and light intensity.

The important features of an instrumentation amplifier are:

- High gain accuracy.
- High CMRR.

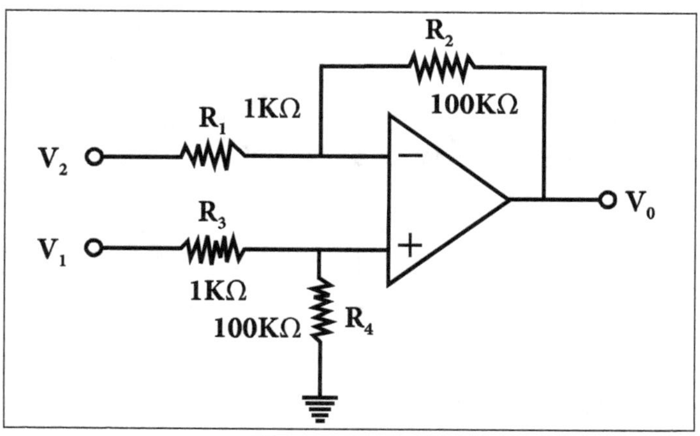

Differential amplifier.

- High gain stability with low temperature coefficient.
- Low AC offset.
- Low output impedance.

There are specially designed op-amps such as μA725 to meet the above stated requirements of good instrumentation amplifier. Monolithic (single chip) instrumentation amplifiers are also available commercially such as AD521, AD524, AD624.

It can be easily seen that the output voltage V_o is given by,

$$V_o = -\frac{R_2}{R_1}V_2 + \left[1+\frac{R_2}{R_1}\right]\left[\frac{R_4}{R_3+R_4}\right]$$

$$V_o = -\frac{R_2}{R_1}V_2 + \frac{1}{1+\frac{R_3}{R_4}}V_1\left(1+\frac{R_2}{R_1}\right) \qquad ...(1)$$

So,

$$V_o = -\frac{R_2}{R_1}\left[V_2 - \frac{1}{1+\frac{R_3}{R_4}}\left(\frac{R_1}{R_2}+1\right)V_1\right] \qquad ...(2)$$

For,

$$\frac{R_1}{R_2} = \frac{R_3}{R_4}$$

We obtain,

$$V_o = \frac{R_2}{R_1}(V_1 - V_2) \qquad ....(3)$$

i.e., the output voltage V_o is the difference of the two input voltages with the gain of $\frac{R_2}{R_1}$.

In the above circuit V_1 has an input impedance = (R_3+R_4 =101kΩ) and the impedance of source V_2 is only R_1 (1kΩ).

This low impedance may load the signal source heavily. This limitation is overcome by the use of voltage follower between each signal input terminal and the difference amplifier.

An improved instrumentation amplifier results as in the below figure.

$$V_2' = R'I + V_2$$

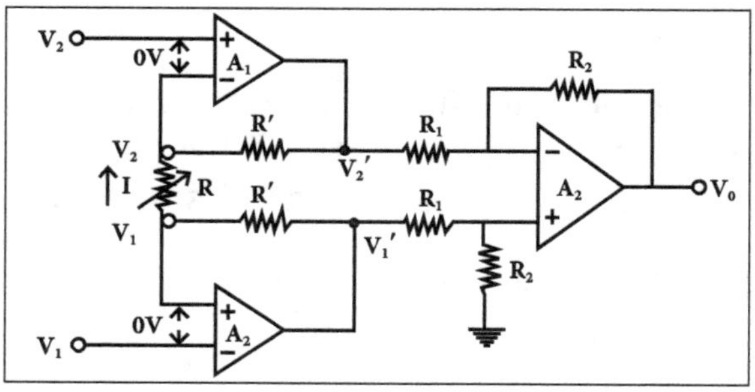

Instrumentation Amplifier.

The Op-amp A_1 and A_2 shown in the above figure are voltage follower (or) buffer circuit act as the input stage for each of the inputs V_1 and V_2.

The Op-amps A_1 and A_2 have differential input voltages at zero.

Case 1: For $V_1 = V_2$, that is, under common mode condition, the voltage across R will be zero.

The voltages at the inverting terminals of the buffers are equal to the input voltages. Since no current flows through R and R' the non-inverter amplifier acts as voltage follower.

So its output $V_2' = V_2$. Similarly, op-amp A_2 acts as voltage follower having output $V_1' = V_1$.

Case 2: For $V_1 \neq V_2$ current flows in R and R' and $[(V_2 - V_1') > (V_2 - V_1)]$.

Therefore, this circuit has more differential gain and CMRR compared to the single op-amp circuit.

The output voltage V_0 can be calculated as follows,

The voltage at the (+) input terminal of op-amp A_3 is

$$\frac{R_2 V_1'}{R_1 + R_2}$$

Using superposition theorem, we have,

$$Vo_2 = \frac{R_2}{R_1}.V_2' \text{ and } V_{01} = \left(1 + \frac{R_2}{R_1}\right) \times \left(V_1' \cdot \frac{R_2}{R_1 + R_2}\right)$$

$$V_0 = V_{0_1} + V_{0_2}$$

$$V_o = \frac{R_2}{R_1} V_2' + \left(1 + \frac{R_2}{R_1}\right)\left(\frac{R_2 V_1'}{R_1 + R_2}\right) \quad \text{...(4)}$$

$$= \frac{R_2}{R_1} V_2' + \frac{V_2'\left(\frac{R_2}{R_1 R_2} \times \frac{R_1 + R_2}{R_1}\right)}{(R_1 + R_2)}$$

$$V_o = \frac{R_2}{R_1}(V_1' - V_2') \quad \text{...(5)}$$

Since, no current flows into op-amp, the current flowing (upwards) in R is $I = \frac{(V_1 - V_2)}{2}$ and passes through the resistor R'.

$$V_1' = R' I + V_1 = \frac{R'}{R}(V_1 - V_2) + V_1 \quad \text{...(6)}$$

$$V_2' = R' I + V_2 = \frac{-R'}{R}(V_1 - V_2) + V_2 \quad \text{...(7)}$$

Putting the values of V_1' and V_2' in equation (5), we obtain,

$$V_o = \frac{R_2}{R_1}\left[\frac{R'}{R}(V_1 - V_2) + V_1 + \frac{R'}{R}(V_1 - V_2) - V_2\right]$$

$$V_o = \frac{R_2}{R_1}\left[2\frac{R'}{R}(V_1 - V_2)(V_1 - V_2)\right]$$

$$V_o = \frac{R_2}{R_1}\left[1 + \frac{2R'}{R}\right](V_1 - V_2)$$

By using a variable resistor R, the gain of this instrumentation amplifier can be varied. The difference gain of this instrumentation amplifier can be varied using variable resistance R.

Since most of the instrumentation systems use the transducer in a bridge circuit, we will consider a simplified differential instrumentation system arrangement using a transducer bridge circuit.

Instrumentation Amplifier using Transducer Bridge

The figure below shows the simplified differential amplifier using Transducer Bridge. The circuit uses a resistance transducer whose resistance changes as a function of the physical quantity to be measured and is denoted by ($R_T \pm \Delta R$),

where,

RT is the resistance of the transducer.

ΔR is the change in resistance.

The bridge is initially balanced by DC supply voltage V_{dc}, so that $V_1 = V_2$. As the physical quantity changes, the resistance R_T of the transducer also changes causing an unbalance in the bridge ($V_1 \neq V_2$).

The differential voltage now gets amplified by the three op-amp differential instrumentation amplifier.

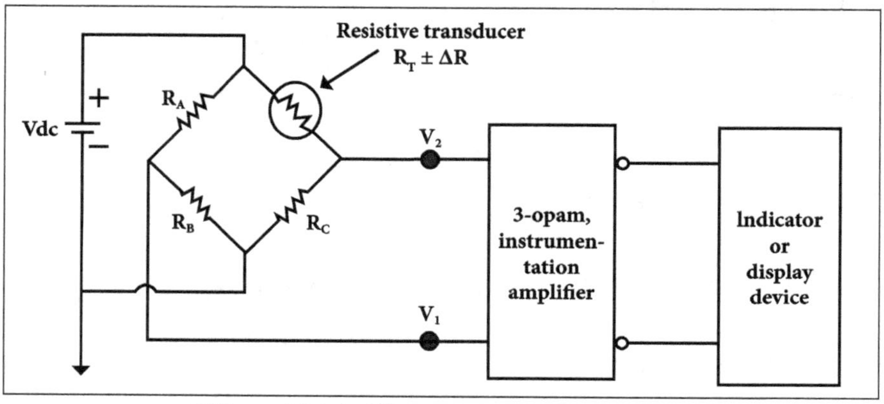

Differential Amplifier Using Transducer Bridge.

1.3.3 AC Amplifier

The op-amp can amplify both the types of signals d.c. and a.c. The op-amp is responding to AC signal is called an AC amplifier. For such an AC amplifier, the low and high frequency limits must be taken care. Similarly, if the AC input signal is along with some d.c. level, it is necessary to use such an AC amplifier, with a coupling capacitor.

For example, In an audio system, there are a number of stages. It is possible to produce a d.c. level in such a system due to thermal drift, component tolerances and changes. To restrict the amplification of such a d.c. level, the coupling capacitors must be used.

The two types of a.c. amplifiers are:

- Inverting AC amplifier.
- Non-inverting AC amplifier.

Inverting AC Amplifier

As the input offset voltage and input offset current can produce voltage at the output, the compensating resistance and network is required. The resistance Rcomp is offset minimizing resistor. The offset voltage compensating network is also shown in the below circuit.

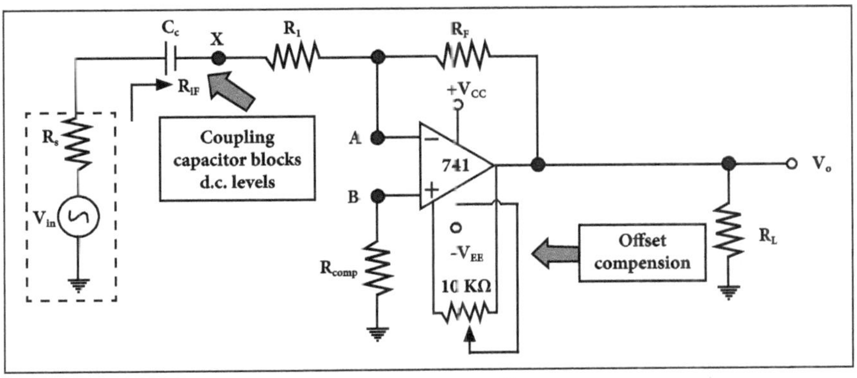

Inverting a.c. Amplifier.

The coupling capacitor Cc blocks any d.c. level existing due to previous stages. The coupling capacitor also provides a low frequency cut-off limit. As the input current of op-amp is zero the node B is at ground potential. Hence the node A is also at virtual ground. The source resistance is R, while R_{iF} is the a.c. input resistance of a.c. amplifier. The voltage at node X i.e. V_x, is amplified by the a.c. inverting amplifier.

Consider the input circuit of the a.c. amplifier as shown in the below figure:

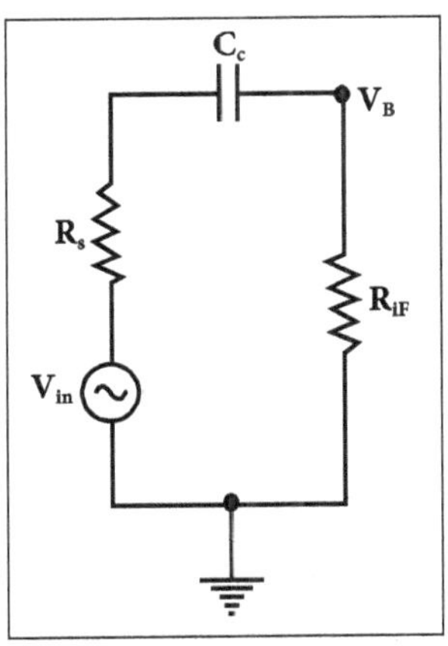

Output circuit.

Using voltage divider rule

$$V_x = \frac{R_{iF}(V_{in})}{R_s + \frac{1}{j\omega C_e + R_{iF}}} \quad ...(1)$$

$$= \frac{j\omega C_e R_{iF}(V_{in})}{(R_s + R_{iF})j\omega C_e + 1}$$

Now,

$$\omega = 2\pi f$$

$$\therefore\ V_x = \frac{j(2\pi f C_e R_{iF})(V_{in})}{j(2\pi f C_e)(R_s + R_{iF}) + 1} \quad ...(2)$$

Now,

$$f_L = \frac{1}{2\pi C_e (R_s + R_{iF})} \quad ...(3)$$

$$\therefore\ V_x = \frac{j(2\pi f C_e R_{iF})(V_{in})}{j\left(\frac{f}{f_L}\right) + 1} \quad ...(4)$$

Where,

f_L = Low Frequency cut-off

The output of the a.c. amplifier is,

$$V_o = -\left(\frac{R_f}{R_1}\right)V_x \quad ...(5)$$

$$\therefore\ V_o = -\left(\frac{R_f}{R_1}\right)\left[\frac{j(2\pi f C_e R_{iF})}{j\left(\frac{f}{f_L}\right) + 1}\right] \quad ...(6)$$

$$\therefore\ \frac{V_o}{V_{in}} = -\left(\frac{R_f}{R_1}\right)\left[\frac{j(2\pi f C_e R_{iF})}{j\left(\frac{f}{f_L}\right) + 1}\right] \quad ...(7)$$

So as the frequency f is changed, the gain also increases. But when $f = f_L$, the denominator becomes 1 + jl whose magnitude is $\sqrt{2}$. Hence its effect on the overall gain is $1/\sqrt{2}$ i.e. 20 log $1/\sqrt{2}$ =- 3.01 dB. Hence the frequency f_L is called break frequency or low cut-off frequency. Hence the frequency response for the V_o/V_{in}, of a.c. amplifier can be shown as in the below figure. For $f > f_L$, the response of the amplifier is fiat in nature.

The high frequency cut-off fH depends on the closed loop gain of the amplifier and unity gain bandwidth of the amplifier. Hence the bandwidth of the amplifier is,

$$BW = f_H - f_L \qquad ...(8)$$

Thus coupling capacitor Cc controls the low frequency limit. The closed loop gain of the a.c. inverting amplifier is still.

$$A_F \cong -\frac{R_f}{R_1} \qquad ...(9)$$

This is because the reactance of Cc is negligible within the bandwidth.

Non-Inverting AC Amplifier

The 100 k Ω pot is used for offset voltage compensation. Rs is the source resistance and R_{iF} is the input a.c. resistance of the amplifier. The input circuit of the non-inverting a.c. amplifier is shown in the below figure. The circuit is same as the input circuit fur the inverting a.c. amplifier. Hence the results obtained for inverting amplifier are equally applicable for non-inverting a.c. amplifier.

$$\therefore\ V_B = \frac{j(2\pi f C_e R_{iF})(V_{in})}{\left(j\frac{f}{f_L}\right)+1} \qquad ...(10)$$

And,

$$f_L = \frac{1}{2\pi C_e (R_s + R_{iF})} \qquad ...(11)$$

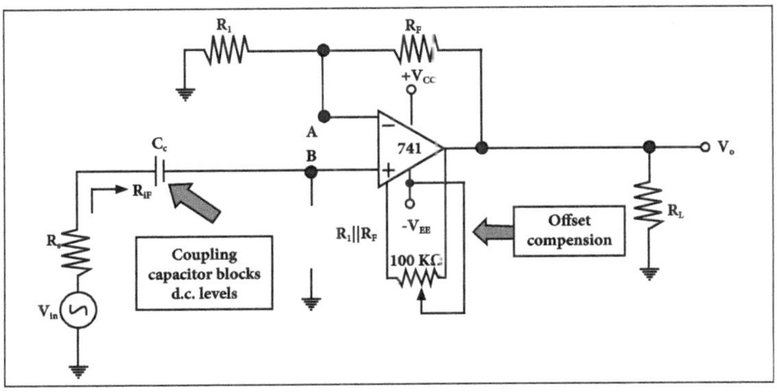

Non-inverting a.c. Amplifier.

While the output is given by,

$$V_o = \left(1+\frac{R_f}{R_1}\right)\left[\frac{j(2\pi f C_e R_{iF})}{\left(j\frac{f}{f_L}\right)+1}\right](V_{in}) \qquad ...(12)$$

Since the reactance Cc is negligible within the bandwidth the closed loop gain of the non-inverting ac amplifier is;

$$A_F \cong 1+\frac{R_f}{R_1} \qquad ...(13)$$

The lower cut off frequency is again controlled by the coupling capacitor Cc.

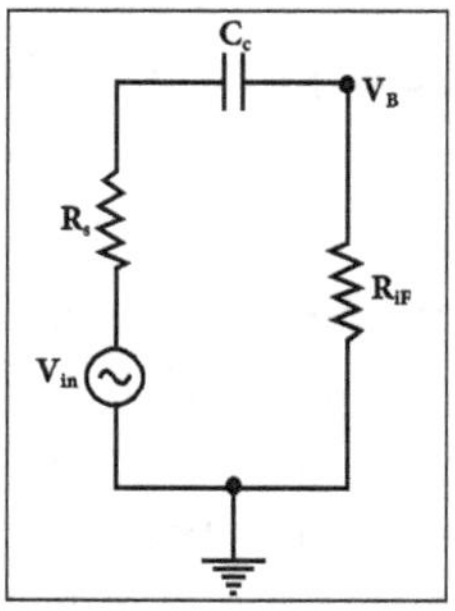

Output circuit.

1.4 Differentiators and Integrators, Comparators and Schmitt Trigger

Differentiators and Integrators

The expression for the output voltage may be determined from KCL.

$$I_c = i_B + i_f$$

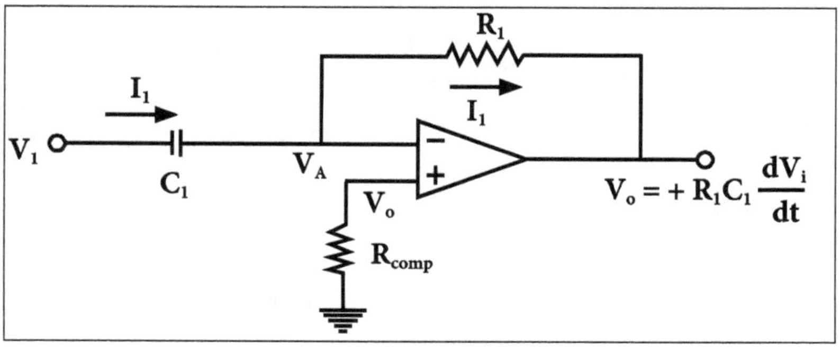

Ideal Differentiator.

$\because\ i_B = 0 \qquad i_c = i_f$

$$C_1 \frac{d(V_1 - V_0)}{dt} = \frac{V_A - V_0}{R_f}$$

$$C_1 \frac{dV_1}{dt} = \frac{-V_o}{R_f}$$

$$V_o = -R_f C_1 \frac{dV_1}{dt}$$

Limitations:

- Noise fluctuations when differentiator generate large noise signals at the output, resulting in a poor signal to noise ratio.
- At high frequencies, the differentiator becomes unstable due to high gain. Therefore, the circuit becomes very sensitive to noise and the stability is affected.

Practical Differentiator

The current at the node A is zero.

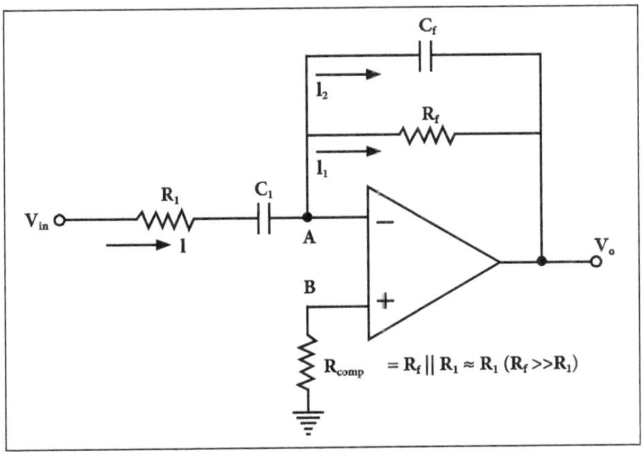

Practical Differentiator.

The input current, $i_c = \dfrac{V_i - V_A}{Z_1}$

$$\frac{V_i}{Z_1} Z_1 \rightarrow R_1 \text{ in series with } C_1$$

$\therefore$ The input impedance, $Z_1 = R_1 + \dfrac{1}{RC_1} = \dfrac{1 + sR_2C_1}{sC_1}$

$$I_C = \frac{sC_1 V_1(s)}{(1+sR_1 C_1)}$$

The current if1 is given by,

$$i_{f1} = \frac{V_A - V_o}{R_f} = \frac{-V_o}{R_f} = \frac{V_o(s)}{R_f}$$

$$i_{f2} = C_f \frac{d(V_A - V_o)}{dt} = -C_f \frac{dV_o}{dt}$$

Taking the Laplace transform,

$$i_C = i_{f1} + i_{f2}$$

$$\frac{sC_1 V_i(s)}{(1+R_1 C_1)} = -\frac{V_o(s)}{R_f} = -sC_f V_o(s)$$

$$\therefore \qquad V_o(s) = \frac{sR_f C_1 V_i(s)}{(1+sR_f C_f)(1+sR_1 C_1)}$$

$$\frac{V_o(s)}{V_i(s)} = \frac{-sR_f C_1}{\left(1+j\frac{f}{f_b}\right)}$$

$$f_{max} = 100\text{Hz}$$

When the value C_1=0.1µf

$$f_a = \frac{1}{2\pi R_f C_1} \quad R_f = \frac{1}{2\pi(100)(10^{-7})}$$

$$R_f = 16\ \text{k}\Omega$$

$$f_b = 10\ f_a = 1000\text{Hz.}$$

$$f_b = \frac{1}{2\pi R_1 C_1} \quad R_1 = \frac{1}{2\pi(100)(10^{-7})}$$

$$= 16\text{k}\Omega$$

As,

$$R_f C_f = R_1 C_1$$

$$C_f = \frac{16\times10^3 \times 0.1\times10^{-6}}{16\times10^3} = 0.1\mu f$$

Integrator

Process of integration is complementary to that of differentiation. A circuit in which the output voltage waveform is the integral of input voltage waveform is the integrator. Such a circuit is determined by using a basic inverting amplifier configuration with the feedback resistor Rf replaced by a capacitor Cf.

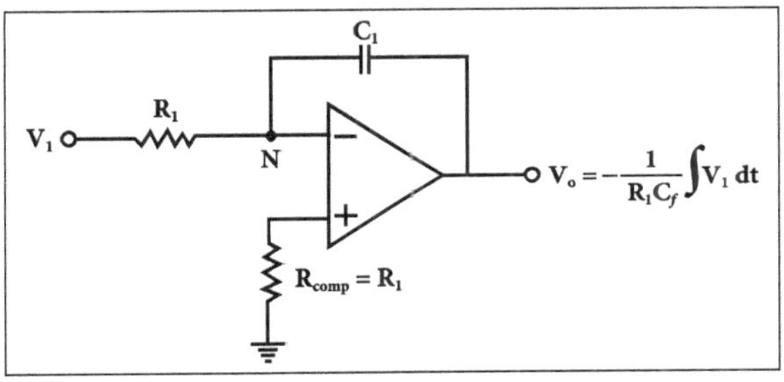

Integrator.

Analysis

The nodal equation at the node N is,

$$\frac{v_i}{R_1} + C_f \cdot \frac{dv_o}{dt} = 0$$

$$C_f \frac{dv_o}{dt} = \frac{-v_i}{R_1}$$

$$\frac{dv_o}{dt} = -\frac{v_i}{C_f R_1}$$

By integrating both sides, we get as,

$$\int_o^t \frac{dv_o}{dt} = -\frac{1}{R_1 C_f}\int_o^t v_i\,dt$$

$$v_o(t) = -\frac{1}{R_1 C_f}\int_o^t v_i(t)dt + v_o(o) \qquad ...(1)$$

Where, v_o (o) is the initial output voltage. Hence, the output is $-\frac{1}{R_1 C_f}$ times the integral of input and $R_1 C_f$ is the time constant of the integrator.

1.4.1 Comparators

A comparator compares two analog signals and produces one bit digital signal.

The comparator output satisfies the following rules:

- When V+ is smaller than V-, the output bit is 0.
- When V+ is larger than V-, the output bit is 1.

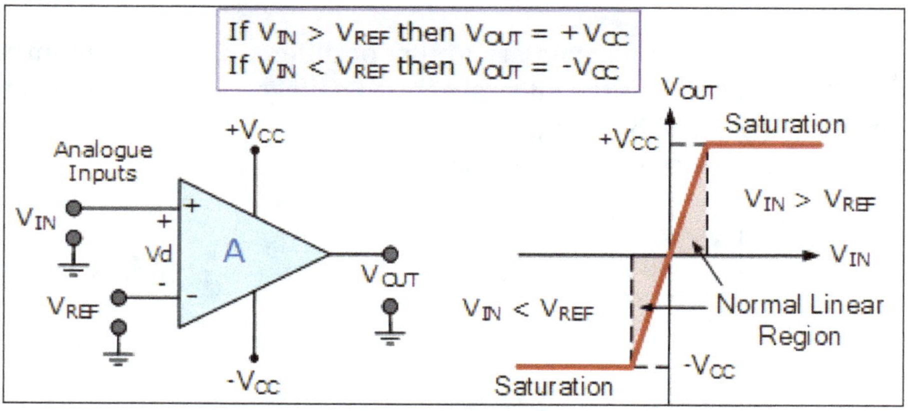

Comparator And Its Characteristic Curve.

The op-amp voltage comparator is a device whose output is dependent on the value of the input voltage, VIN with respect to a particular DC voltage level as the output is HIGH, if the voltage on the non-inverting input is greater than voltage on inverting input.

And LOW, if the non-inverting input is less than the inverting input voltage. The condition is true regardless of whether the input signal is connected to the inverting or non-inverting input of the comparator.

The basic op-amp comparator produces positive or negative voltage output by means of comparing its input voltage against some of the preset DC reference voltage.

Voltage Comparator Circuit

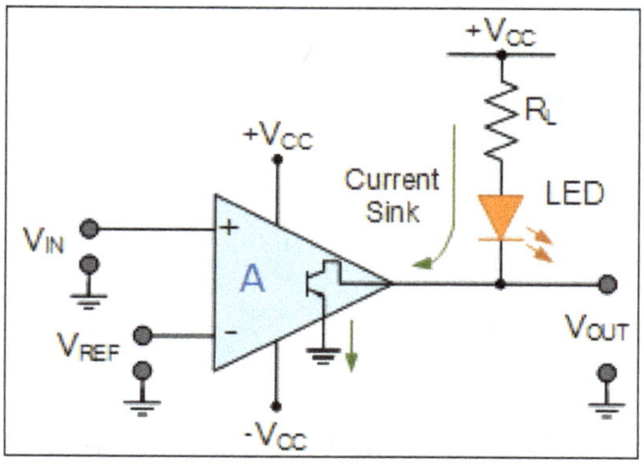

Voltage comparator circuit.

Here, the open collector output from voltage comparator is connected to voltage source via a single pull-up resistor that pulls the single output high to the power supply.

When the output switch is HIGH, it creates a high impedance path. Hence, no current flows such that $V_{OUT} = V_{cc}$.

1.4.2 Schmitt Trigger

If a positive feedback is added to the comparator circuit, gain can be increased greatly. The below figure shows a regenerative comparator:

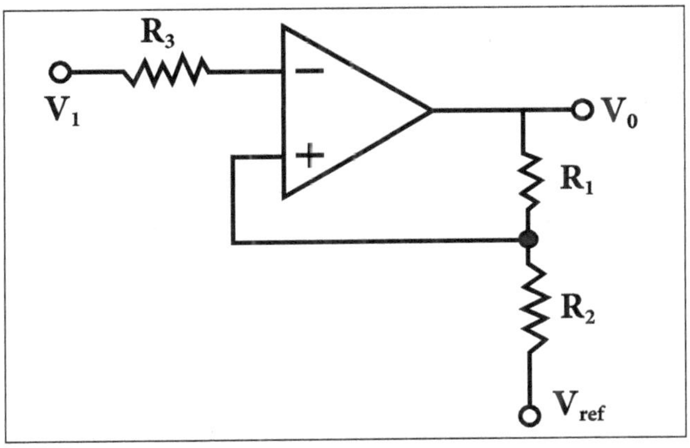

Inverting Schmitt Trigger.

The above given circuit is also known as Schmitt trigger or Squaring circuit. The input voltage is applied to the (–) input terminal and feedback voltage to the (+)ve input terminal. The input voltage K triggers the output Vo every time when it exceeds certain voltage levels.

These voltage levels are called upper threshold voltage (V_{UT}) and lower threshold voltage (V_{LT}). The hysteresis width is the difference between these two threshold voltages V_{UT} - V_{LT}. These threshold voltages are calculated as follows:

Suppose the output $V_o = +V_{sat}$. The voltage at (+) input terminal will be,

$$V_{ref} + \frac{R_2}{R_1 + R_2} \cdot (V_{sat} - V_{ref}) = V_{UT} \qquad ...(1)$$

This voltage is called upper threshold voltage V_{UT}. As long as V_i is less than V_{UT}, the output V_o remains constant at $+V_{sat}$. When Vi is just greater than V_{UT}, the output regeneratively switches to $-V_{sat}$ and remains at this level as long as $V_i > V_{UT}$ as shown in the below figure (b).

For Vo = $-V_{sat}$, the voltage at the (+) input terminal is,

$$V_{ref} - \frac{R_2}{R_1 + R_2} \cdot (V_{sat} + V_{ref}) = V_{LT} \qquad ...(2)$$

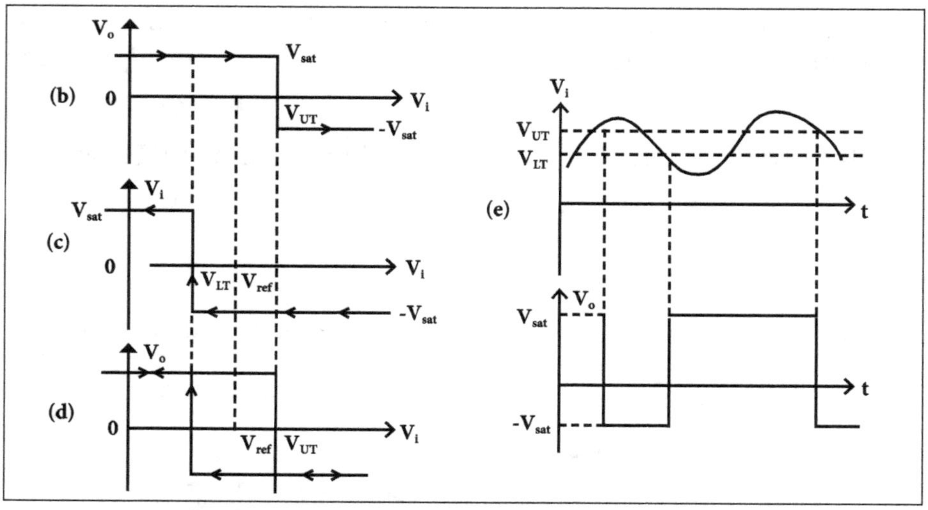

Transfer characteristics for: (b) Vi increasing, (c) Vi decreasing, (d) Composite Input-output curve, (e) Schmitt trigger used as a squarer.

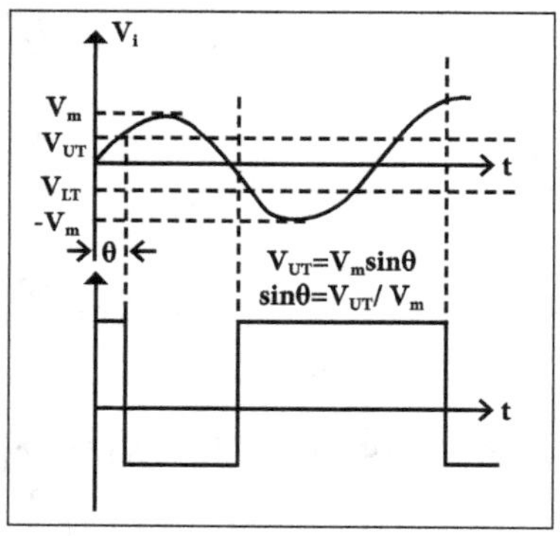

Shift θ in the output waveform for $V_{UT} - V_{LT}$.

This voltage is referred as lower threshold voltage V_{LT}. The input voltage V_i must become lesser than V_{LT} in order to cause V_o to switch from $-V_{sat}$ to $+V_{sat}$. A regenerative transition takes place and the output V_o returns from $-V_{sat}$ to $+V_{sat}$ almost instantaneously. The complete transfer characteristics are shown in the above figure.

Note that VLT < VUT and the difference between these two voltages is the hysteresis width VH and can be written as,

$$V_H = V_{UT} - V_{LT} = \frac{2R_2 V_{sat}}{R_1 + R_2} \qquad ...(3)$$

Because of the hysteresis, the circuit triggers at a higher voltage for increasing signals than for decreasing ones. Further, note that if peak-to-peak input signal Vi were smaller than VH, then the Schmitt trigger circuit having responded at a threshold voltage by a transition in one direction would never reset itself. That is once the output has jumped to say +Vsat, it would remain at this level and never return to −Vsat.

It may be seen from equation $V_H = V_{UT} - V_{LT} = \frac{2R_2 V_{sat}}{R_1 + R_2}$ that hysteresis width VH is independent of Vref. The resistor R3 in the above figure (a) is chosen equal to R1 R2 to compensate for the input bias chosen.

A non-inverting Schmitt trigger is obtained as Vi and Vref are interchanged in the above figure. The most important application of Schmitt trigger circuit is to convert a very slowly varying input voltage into a square wave output as shown in the above figure.

In the circuit, Vref is chosen as zero as it follows from equation (1) and (2), so that

$$V_{UT} = -V_{LT} = \frac{R_2 V_{sat}}{R_1 + R_2} \quad \text{...(4)}$$

If the sinusoidal input of frequency f = 1/T is applied to such a comparator, a symmetrical square wave is obtained at the output. The vertical edge of the output waveform however will not occur at the time the sine wave passes through zero, but it is shifted in phase by θ where $\sin\theta = V_{UT}/V_m$ and V_m is the peak sinusoidal voltage.

When the output is +Vsat then reference voltage V_{ref} is given by,

$$V_{ref} = \frac{R_1}{(R_1 + R_2)} * V_{sat} = (+\beta V_{sat})$$

If V_{in} is less than V_{ref} output will remain $+V_{sat}$.

When input v_{in} exceeds $V_{ref} = +V_{sat}$ the output switches from $+V_{sat}$ to $-V_{sat}$. Then the reference voltage is given by,

$$V_{ref} = \frac{-R_2}{(R_1 + R_2)} * V_{sat} = (-\beta V_{sat})$$

The output will remain $-V_{sat}$ as long as $V_{in} > V_{ref}$.

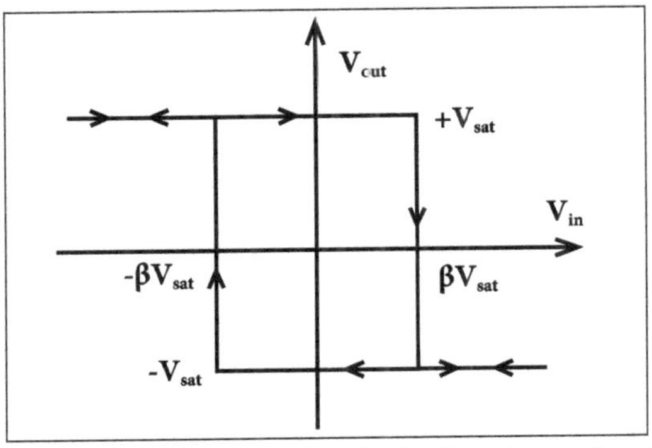

The transfer characteristic of Schmitt trigger circuit.

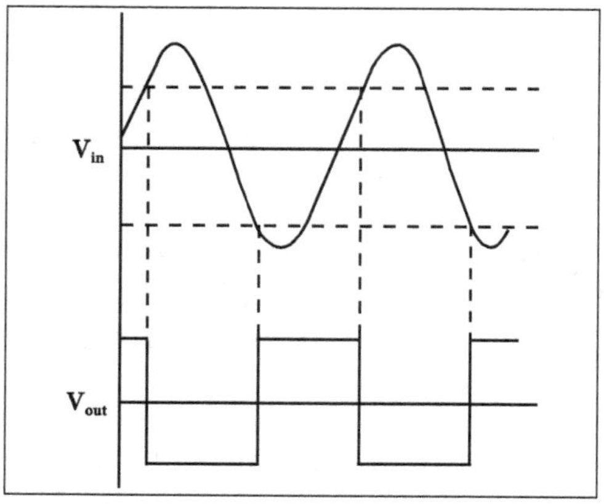

The output for a sinusoidal wave.

If $V_{in} < V_{ref}$ i.e. v_{in} becomes more negative than $-V_{sat}$ then again output switches to $+V_{sat}$ and so on. If the input is different than sine even then the output will be determined in a same way.

Special purpose Schmitt triggers are commercially available,

e.g. T_1–13, T_1–14 chips

CMOS Schmitt triggers offers the advantage of high input impedance and low power consumption.

1.5 Voltage Regulators, Features of 723 Regulator and Three Terminal Voltage Regulators

Voltage regulator, any electrical or electronic device that maintains the voltage of a power source within acceptable limits. The voltage regulator is needed to keep voltages within the prescribed range that can be tolerated by the electrical equipment using that voltage. Such a device is widely used in motor vehicles of all types to match the output voltage of the generator to the electrical load and to the charging requirements of the battery. Voltage regulators are also used in electronic equipment in which excessive variations in voltage would be detrimental.

In motor vehicles, voltage regulators rapidly switch from one to another of three circuit states by means of a spring-loaded, double-pole switch. At low speeds, some current from the generator is used to boost the generator's magnetic field, thereby increasing voltage output.

At higher speeds, resistance is inserted into the generator-field circuit so that its voltage and current are moderated. At still higher speeds, the circuit is switched off, lowering the magnetic field. The regulator switching rate is usually 50 to 200 times per second.

Electronic voltage regulators utilize solid-state semiconductor devices to smooth out variations in the flow of current. In most cases, they operate as variable resistances, that is, resistance decreases when the electrical load is heavy and increases when the load is lighter.

Voltage regulators perform the same function in large-scale power-distribution systems as they do with motor vehicles and other machines, they minimize variations in voltage in order to protect the equipment using the electricity. In power-distribution systems the regulators are either in the substations or on the feeder lines themselves.

Two types of regulators are used: step regulators, in which switches regulate the current supply and induction regulators, in which an induction motor supplies a secondary, continually adjusted voltage to even out current variations in the feeder line.

IC Voltage Regulators

Depending upon the control element is connected in the regulator circuit the regulators are basically classified as,

- Series Voltage Regulator.
- Shunt Voltage Regulator.

Series Voltage Regulator

The control element is connected in series with the load. The unregulated DC voltage is the input to the circuit. Control element controls the amount of the input voltage.

The sampling circuit provides the necessary feedback signal. The comparator circuit compares the feedback with the reference voltage to generate the appropriate control signal.

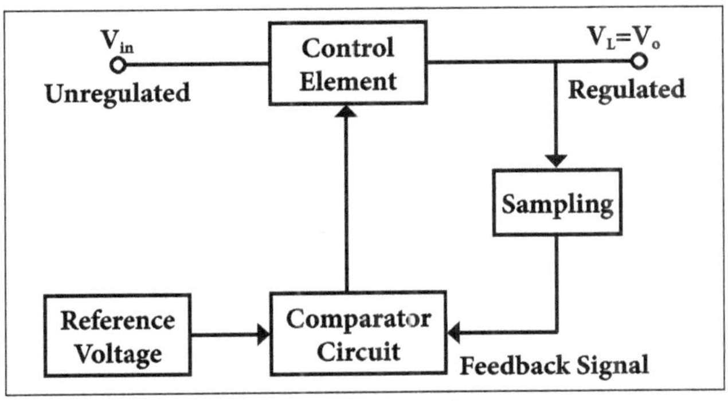

Series Voltage Regulator.

When load voltage tries to increase, the comparator generates a control signal based on the feedback information. Control signal causes the control element to decrease the amount of the output voltage. Output voltage is maintained as constant.

Shunt Voltage Regulator

Shunt regulators are simpler than series regulators. It consists of voltage-reference diodes and are used in very low-powered circuits. Shunt regulators are very common in voltage reference circuits.

There are several applications for shunt regulators including:

- Low Output Voltage Switching Power Supplies.
- Current Source and Sink Circuits.
- Error Amplifiers.
- Adjustable Voltage or Current Linear and Switching Power Supplies.
- Voltage Monitoring.
- Analog and Digital Circuits Requiring Precision References.
- Precision current limiters.

1.5.1 Features of 723 Regulators

The important features of IC 723 regulators are given below:

- It has small in size and lower in cost.
- It has choice of supply voltage.
- It operates in positive or negative supply operation.
- Low standby current gain.
- Wide variety of applications such as series, shunt, switching and floating regulators.
- Relative simplicity with power supply can be designed.
- Very low temperature drift and high ripple rejection.
- Built in fold back current limiting.
- Input voltage is maximum 40 V.
- Output voltage adjustable from 2 V to 37 V.
- Output current up to 150 mA without external pass transistor.
- Load and line regulations of 0.03%.

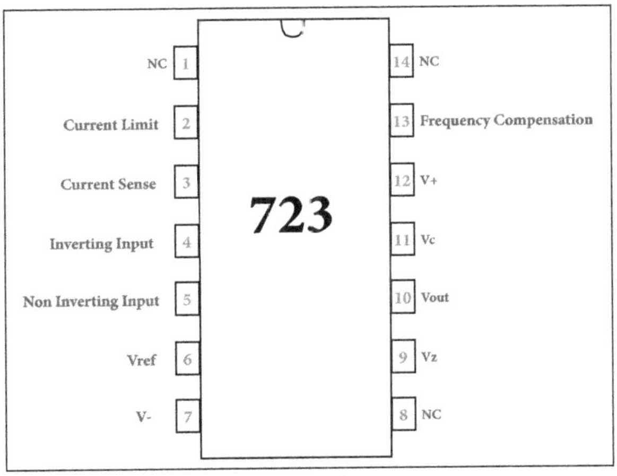

IC 723 regulators.

1.5.2 Three Terminal Voltage Regulators

An IC based voltage regulator can be classified in different ways. A common type of classification is 3 terminal voltage regulator and 5 or multi terminal voltage regulator. Another popular way of classifying IC voltage regulators is by identifying them as linear voltage regulator & switching voltage regulator.

There is a third set of classification as:

- Fixed voltage regulators (positive & negative).
- Adjustable voltage regulators (positive & negative).
- Switching regulators.

In the third classification, fixed & adjustable regulators are basically versions of linear voltage regulators.

Block Diagram of three Terminals IC based Voltage Regulator

We have given below the block diagram of a 3 terminal IC based voltage regulator:

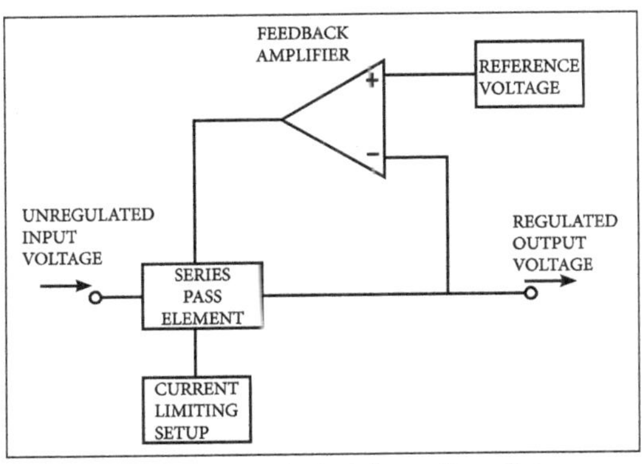

Block diagram of IC Voltage Regulator.

Fixed Voltage Regulators

These regulators provide a constant output voltage. A popular example is the 7805 IC which provides a constant 5 volts output. A fixed voltage regulator can be a positive voltage regulator or a negative voltage regulator. A positive voltage regulator provides with constant positive output voltage. All those IC's in the 78XX series are fixed positive voltage regulators. In the IC nomenclature – 78XX; the part XX denotes the regulated output voltage the IC is designed. For examples: 7805, 7806, and 7809 etc.

A negative fixed voltage regulator is same as the positive fixed voltage regulator in design, construction & operation. The only difference is in the polarity of output voltages. These IC's are designed to provide a negative output voltage. Example: 7905, 7906 and all those IC's in the 79XX series.

Adjustable Voltage Regulator

An adjustable voltage regulator is a kind of regulator whose regulated output voltage can be varied over a range. There are two variations of the same known as positive adjustable voltage regulator and negative adjustable regulator. LM317 is a classic example of positive adjustable voltage regulator, whose output voltage can be varied over a range of 1.2 volts to 57 volts. LM337 is an example of negative adjustable voltage regulator. LM337 is actually a complement of LM317 which are similar in operation & design; with the only difference being polarity of regulated output voltage.

There may be certain conditions where a variable voltage may be required. The connection diagram is shown in the below figure.

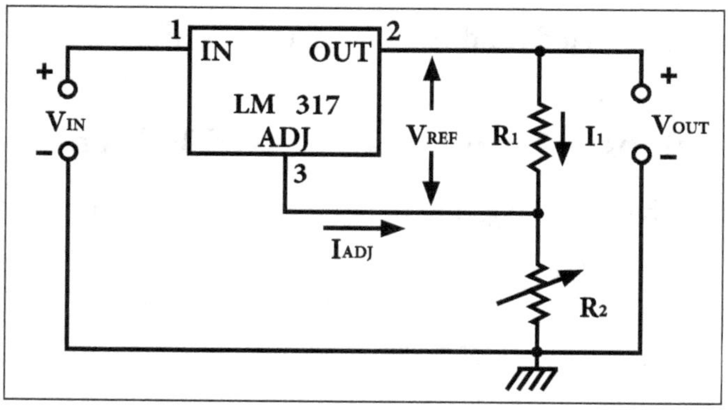

Adjustable Voltage Regulator using LM317.

The resistors R1 and R2 determine the output voltage V_{out}. The resistor R_2 is adjusted to get the output voltage range between 1.2 volts to 57 volts. The output voltage that is required can be calculated using the below equation:

$$V_{out} = V_{ref}(1+R_2/R_1) + I_{adj}R_2$$

In this circuit, the value of $V_{re}f$ is the reference voltage between the adjustment terminals and the output taken as 1.25 Volt.

The value of I_{adj} will be very small and will also have a constant value. Thus, the above equation can be rewritten as,

$$V_{out} = 1.25\ (1 + R_2/R_1)$$

In the above equation, due to the small value of Iadj, the drop due to R2 is neglected.

The load regulation is 0.1 percent while the line regulation is 0.01 percent per volt. This means that the output voltage varies only 0.01 percent for each volt of input voltage. The ripple rejection is 80 db, equivalent to 10,000.

The LM 337 series of adjustable voltage regulators is a complement to the LM 317 series devices. The negative adjustable voltage regulators are available in the same voltage and current options as the LM 317 devices.

LM340 Series Voltage Regulator

The voltage regulator using LM340 IC is the most used voltage regulator IC. As shown in the block diagram below, the built-in reference voltage V_{ref} drives the non-inverting input of the operational amplifier. There are many stages of voltage gain for the op-amp used here.

This high gain helps the op-amp to make the error voltage between the inverting and non-inverting terminals to be almost zero. Thus, the inverting input terminal value will also be the same as the non-inverting terminal, Vref. Thus, the current flowing through the potential divider can be written as,

$$I = V_{ref} / R_2$$

The resistor R_2 shown in the above figure is not an external component connected to the IC, but an internal resistor that is built inside the IC during manufacture. Due to the conditions above, the same current flows through R_1. Thus, the output voltage can be written as,

$$V_{out} = V_{ref} / R_2 \left(R_1 + R_2\right)$$

This shows that the output of the regulator can be controlled by putting desired values for R_1 and R_2.

The IC has a series pass transistor that can handle more than 1.5 A of load current provided that enough heat sinking is provided along with it.

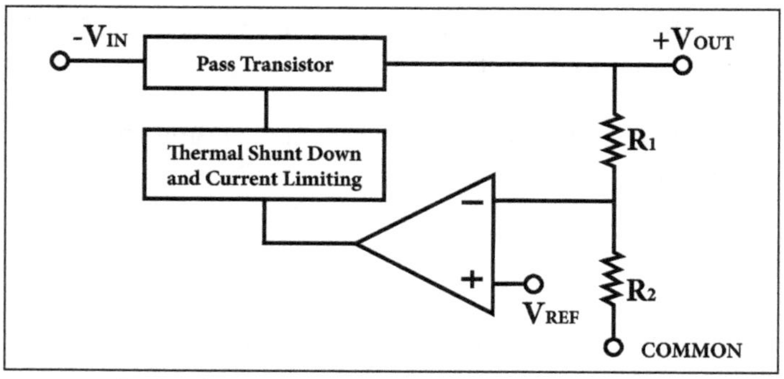

3 Terminal Voltage Regulator IC – Block Diagram – LM340.

Like other IC's, this IC also has thermal shutdown and current limiting options. Thermal shutdown is a feature that will turn off the IC as soon as the internal temperature of the IC rises above its preset value. This rise in temperature may mostly be due to excessive external voltage, ambient temperature or even heat sinking. The preset cut-off temperature value for LM340 IC is 175° C. Because of thermal shutdown and current limiting, devices in the LM 340 series are almost indestructible.

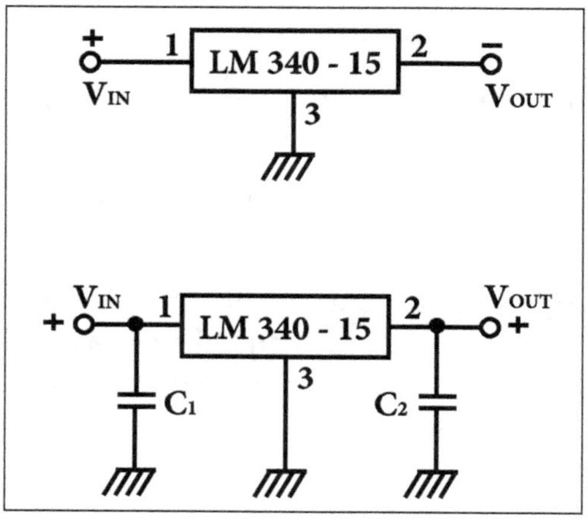

LM 340 – 15 Circuits.

The above figure shows the application of LM340 IC as a voltage regulator. Pins 1, 2, and 3 are the input, output and ground.

When there is quite a distance (in cms) from the IC to the filter capacitor of the unregulated power supply, there may occur unwanted oscillations within the IC due to lead inductances within the circuit. In order to remove this unwanted oscillation, the capacitor C1 has to be placed as shown in the above circuit.

Capacitor C2 is sometimes used to improve the transient response of the circuit.

Any device in the LM 340 series needs a minimum input voltage at least 2 to 3 V greater than the regulated output voltage. Otherwise, it will stop regulating. Furthermore, there is a maximum input voltage because of excessive power dissipation.

Switching Voltage Regulators

Switching regulators differ in design, construction and operation compared to linear voltage regulators. In switching regulators, the output voltage is regulated by controlling the switching time of feedback circuitry (which includes a reference voltage); that is by adjusting the duty cycle.

The regulators which have been shown below are all linear voltage regulators which need a series pass transistor to regulate in the active region. Though, they are chosen for different purposes, they do have a disadvantage of high power dissipation of the series pass transistor. The main reason is because the series pass transistor is used as a Class A power amplifier which is known for its low efficiency and power loss in the form of heat.

The series pass resistor will have to carry a heavy load when the load current increases. This cause the series pass transistor to get bulky with a bulkier heat sink. This in turn increases the overall cost as well. Such linear regulators will also need a step-down transformer which again increases the size of the overall circuit. The large ripples produced by the circuit should be eliminated and this requires large sized filter capacitors.

The basic switching regulator is designed to work in three configurations. Their circuit diagrams and explanation are given below:

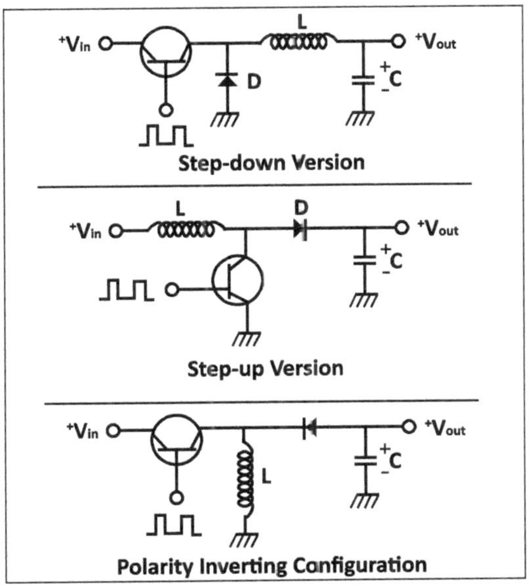

Switching Voltage Regulator – Types.

All these problems can be solved by using a switching voltage regulator. The whole operation is completely different when compared to a linear voltage regulator. Here, the series pass transistor is not used as an amplifier, but as a switch. That is, instead of the transistor working in the active region, it is made to switch between the saturation region or cut-off region. Thus, the power dissipation is lessened and thus can carry heavy load currents at low voltages with less bulky heat sinks. Thus, this regulator finds its extensive use in personal computers.

Step-down Switching Regulator

As shown in the figure above, the rectangular pulses are given to the base of the transistor. During each cycle of the pulse the transistor changes between saturation and cut-off. This creates a rectangular voltage at the input to LC filter. The ac components of the input voltage to the filter is blocked and the dc component is allowed to pass through the filter. As the transistor keeps switching, the average value will always be lesser than the input voltage. This is why we call it the step-down switching regulator.

Step-up Switching Regulator

As shown in the figure above, the rectangular pulses are given to the base of the transistor. During each cycle of the pulse the transistor changes between saturation and cut-off. When the transistor is saturated the current flows through the inductor.

When the transistor switches to cut-off a large voltage will be induced across the inductor coil because of the sudden collapse of the magnetic field around it. Thus, the current keeps flowing in the same direction. The circuit is called a step-up switching regulator because the voltage induced by the inductor will be larger than the input voltage.

Polarity Inverting Switching Regulatory

As shown in the figure above, when the transistor is saturated, the current flows through the inductor. When the transistor switches to cut-off a large voltage will be induced across the inductor coil because of the sudden collapse of the magnetic field around it.

Thus, the current keeps flowing in the same direction. Since the transistor is cut-off, the only path is through the capacitor. If the direction of charging current through the capacitor is checked, output voltage is found to be negative.

Simple Switching Regulator

A simple switching regulator is designed with the mix of circuits. The working starts with the relaxation oscillator which generates a square wave. The frequency of the square wave is decided by the values of R_5 and C_3. The square wave is given as input to the integrator and produces an output triangular wave. This is given as input to the positive terminal of the triangle to pulse converter. The output pulse will then drive the base of the pass transistor. The duty cycle of these pulses will determine the output voltage.

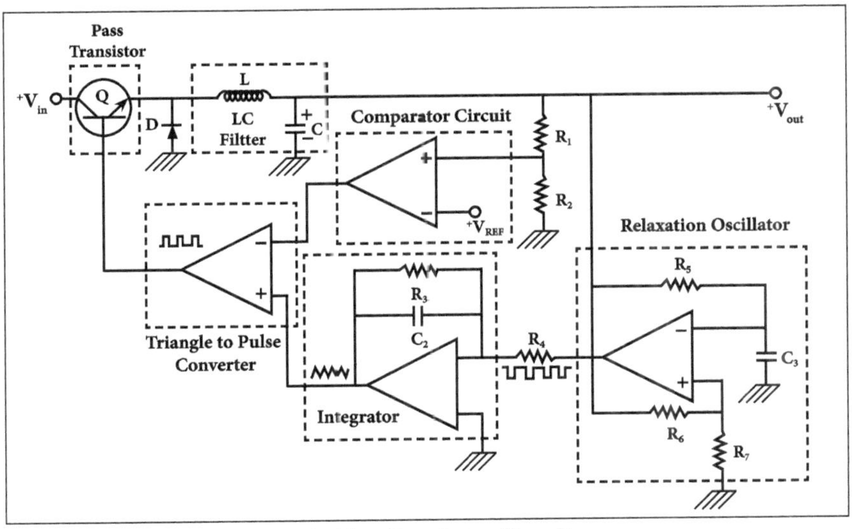

Switching Regulator Circuit.

The duty cycle D is the ratio of the ON time W to the time period T. By controlling the duty cycle of the pulse generator, the duty cycle of the input voltage to the LC filter is controlled. The output of the LC filter is a dc voltage with only a small ripple. This output,

$$V_{out} = DV_{in}$$

Since D can vary from 0 to 1, V_{out} can vary from 0 to V_{in}.

A simple voltage divider is used to sample the output of the LC filter. This voltage is then feedback to the comparator circuit where a reference voltage V_{ref} is compared with the output voltage and passed on to the negative input of the triangle to pulse converter.

When the output voltage increases, the comparator circuit produces a higher output voltage and thus the inverting input of the triangle to pulse converter will have high value. This will narrow down the pulses at the base input of the pass transistor. Since the duty cycle is lower, the filtered output voltage is less, which tends to cancel almost all the original increase in the output voltage. It means that any attempted increase in output voltage generates a negative feedback voltage that almost eliminates the original increase. Reverse happens, the output voltage fall.

There is enough open-loop gain in the system to ensure a well-regulated output voltage. Since the error voltage to the comparator is near zero, the voltage across R_2 is approximately equal to V_{ref}. So the current through resistor R_2 is,

$$I = V_{ref}/R_2$$

This current flows through R_1, which means the output voltage is,

$$V_{out} = V_{ref}(R_1 + R_2)$$

Switching regulators are available in different configurations like the fly back configuration, feed-forward, push-pull and non-isolated single-ended or single-polarity types.

Some of the common IC's used as switching regulators in the above said configurations are the LM 1577/LM 2577 IC's.

2

Filters and Waveform Generators

2.1 Active Filters, Band Pass, Band Reject and all Pass Filters

The filter is an electric network in any circuit theory that is used to change either phase or amplitude of signal characteristics with respect to its frequency. Ideally, this will not include any new frequency to the input nor it will alter the frequency component of that signal.

An active filter utilizes an operational amplifiers along with various electronic components like resistors, capacitors for the filtering. Op-amps are used to allow easily to make many types of active filters.

An amplifier stops the load impedance from affecting the filter characteristics. The form of the response, duality factor and tuned frequency can often a set with inexpensive variable resistors. In these filter circuits we can alter one parameter without damaging the other. Since their basic return principles were projected around 1970, a lot of research has been done with these filters and their realistic applications.

Types of Active Filters

The most common types of active filters are classified into:

Low Pass Filter

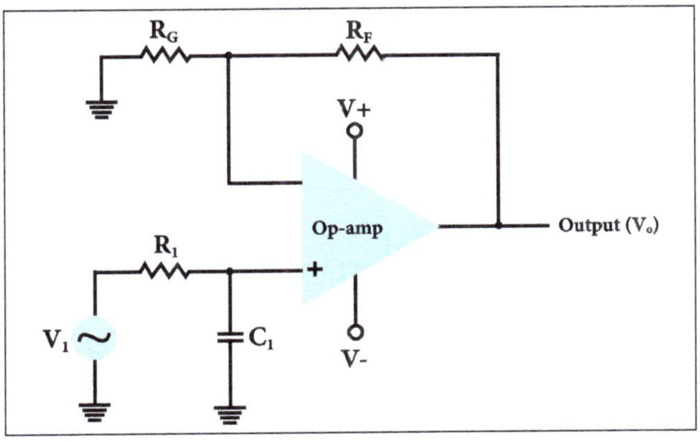

Low pass filter.

A low pass filter is used in circuits that only allow low frequencies to pass through. It is often used to block high frequencies and AC current in a circuit. Given below is a sample low pass filter circuit using op-amp. Ideally, the frequency output of a low pass filter is like this.

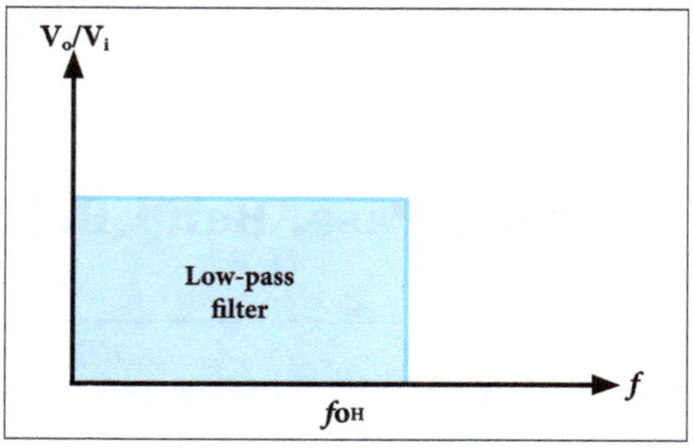

V_O/V_i vs f_{OH}

But this is not the case with real circuits. As there is a slight attenuation.

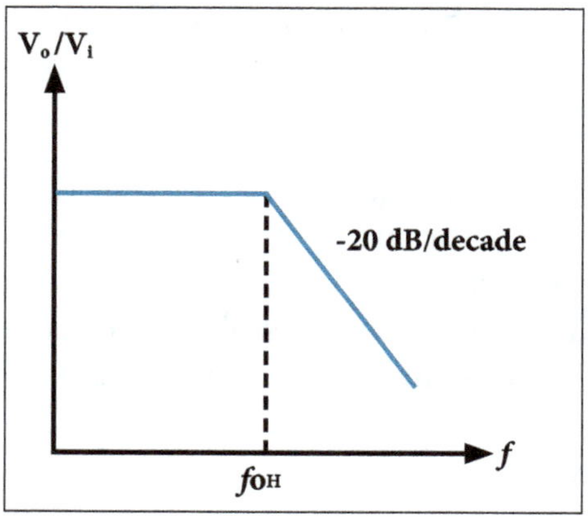

V_O/V_i vs f_{OH}

This attenuation can be further minimized by adding multiple stages as below:

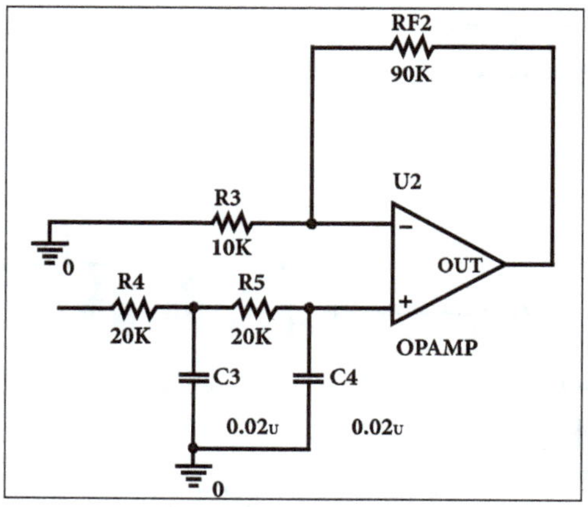

Low Pass Filter Circuit Diagram.

Low pass filter is popular with speakers to block high pitches, some electric guitars and radio transmitters.

High Pass Filter

A high pass filter is used in circuits that only require high frequencies to operate. It blocks most low frequencies & DC component. Given below is a sample high pass filter circuit using op-amp.

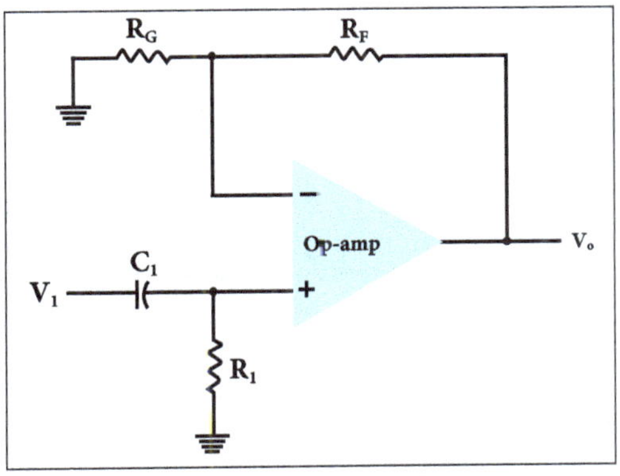

High pass filter - circuit diagram.

Ideally, the frequency output of a high pass filter is like this.

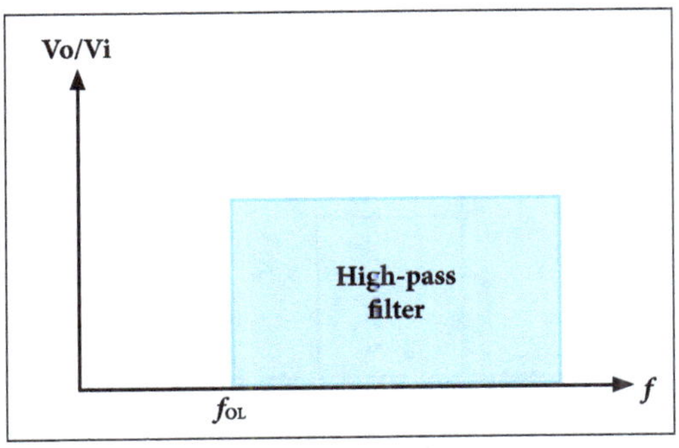

V_o/V_i vs f_{OL}

But this is not the case with practical circuits. As there is a slight attenuation.

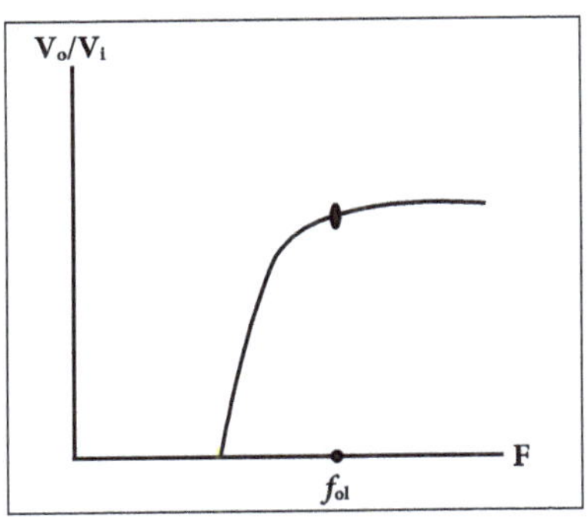

V_o/V_i vs f_{ol}

This attenuation can be further minimized by adding multiple stages as with low pass filter.

Band Pass Filter

A band pass filter is a combination of a high pass and a low pass filter. It allows only a select range of frequencies to pass through. It is designed such a way that the cut off frequency of the low pass filter is higher than the cut off frequency of the high pass filter, hence allowing only a select range of the frequencies to pass through. Presented here is a sample band pass filter circuit using op-amp.

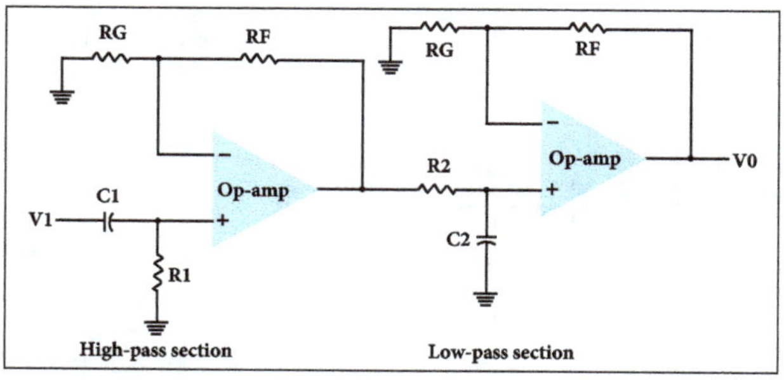

Band Pass Filter.

Ideally, frequency output of a band pass filter is like this,

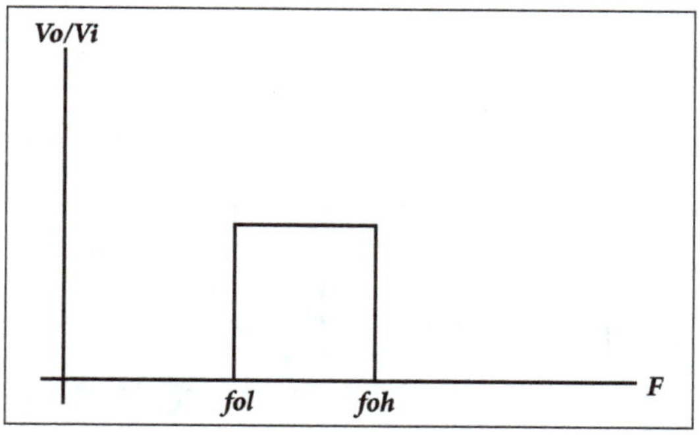

Frequency Output of a Band Pass Filter.

But this is not the case with practical circuits. As there is a slight attenuation.

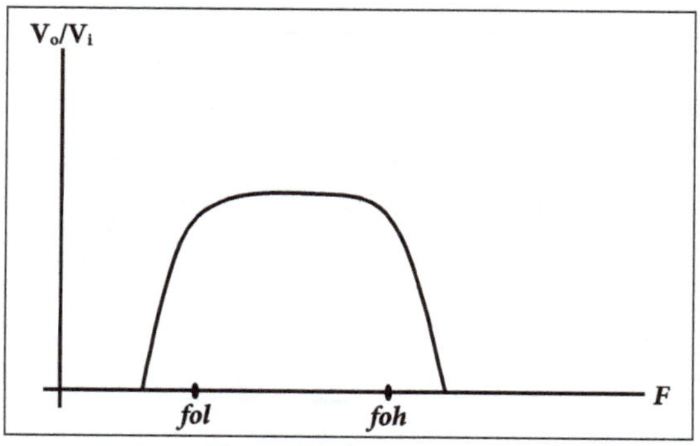

Band pass filter.

Band pass filter is popularly with speakers to block high pitches, some electric guitars and radio transmitters.

Band Stop Filter

A band stop filter also known as notch filter is used in circuits that block only a select range of frequencies and allows others to pass through. It would be an inverse of the band pass filter, and can be created by using the same input at a high pass and a low pass filter.

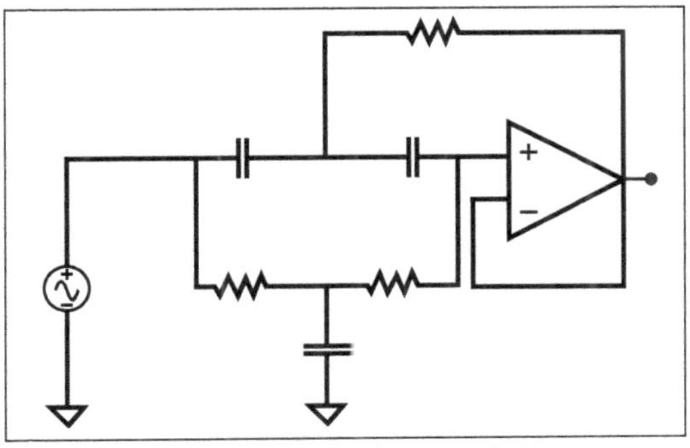

Circuits for Band stop Filter.

There are various kinds of filters are available, but most of the applications can be resolved with these implementations.

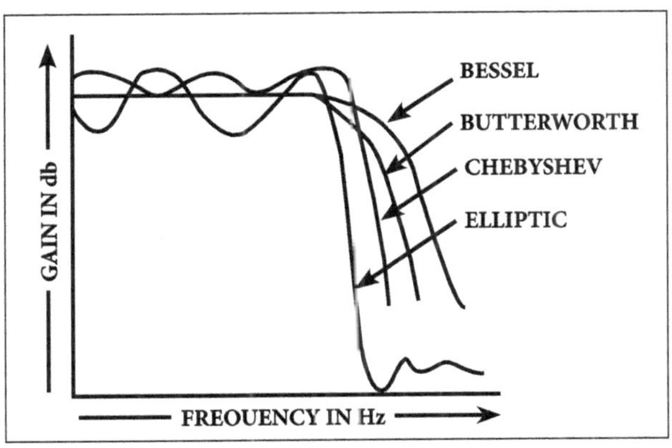

Types of Active Filters.

Advantages of Active Filters

The active filters have the following advantages over the passive filters:

- All the elements along with op-amp can be used in the integrated form. Hence, there is reduction is size and weight.
- The inductors are absent in the active filters hence the modem active filters are more economical.
- Due to flexibility in gain and frequency adjustments, the active filters can be easily tuned.

- The op-amp gain can be easily controlled in the closed loop fashion hence active filter input signals is not attenuated.
- Due to availability of modem ICs, variety of cheaper op-amps are available.
- The op-amp has high input impedance and low output impedance hence the active filters using op-amp do not cause loading of the source or load.
- In large quantities, the cost of the integrated circuit can be much lower than its equivalent passive network.
- Active filters can be realized under number of class of functions such as Butterworth, Thomson, Chebyshev, Cauer etc.
- The response is improved as compared to passive filters due to ready availability of high quality components.
- Active filters can provide voltage gains in contrast the passive filters often show a significant voltage loss.
- The design procedure is simpler than that for the passive filters.

In spite of the above advantages, the active filters also have certain limitations. The finite bandwidth of the active devices places a limit on the highest frequency of operation. For most of the applications, with present day technology, a reasonably good filter performance can be achieved approximately up to 500 kHz. As against this passive filters can be used up to 500 MHz.

Another important factor is the sensitivity. The active elements are much more sensitive to the temperature and the environmental changes than the passive elements. Hence due to changes in the parameters due to the environmental changes, the active filter performance deviates from its ideal response.

The requirement of d.c. power supply is another disadvantage of the active filters. A passive filters do not require the d.c. supply.

2.1.1 Characteristics of Band Pass Band Reject and All Pass Filters

The band-pass filter has a gain response with a frequency range from ω_{C1} to ω_{C2}. Any input that has frequencies between ω_{C1} and ω_{C2} gets a pass and anything outside this range gets attenuated or rejected.

The input signal of the filter shown here has equal amplitude at frequencies ω_1, ω_2 and ω_3. After passing through the band-pass filter, the output amplitudes at ω_1 and ω_3 are significantly decreased because they fall outside the desired frequency range, while the frequency at ω_2 is within the desired range, so its signal amplitude passes through unaffected.

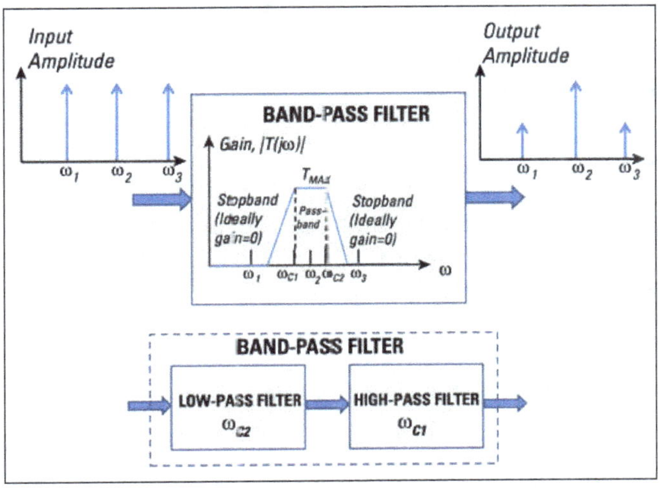

Band pass.

We can think of the band-pass filter as a series or cascaded connection of the low-pass filter with frequency ω_{C2} and a high-pass filter with frequency ω_{C1}.

The cascade connection of the low-pass filter and high-pass filter forms a band-pass filter. The order of the filters doesn't matter.

Band Reject (Eliminated)

The band elimination filter is also called band rejection filter or band stop filter. Its characteristics is exactly opposite to that of band pass filter. There are two pass bands while one stop band. The stop band is between the two frequencies fL and fH.

The two ranges $0 < f < f_1$ and $f_H < f < \infty$ are the two pass bands. The frequency response characteristics of band elimination filter is shown in the below figure. The below figure(a) shows the ideal response while the below figure(b) shows the practical response.

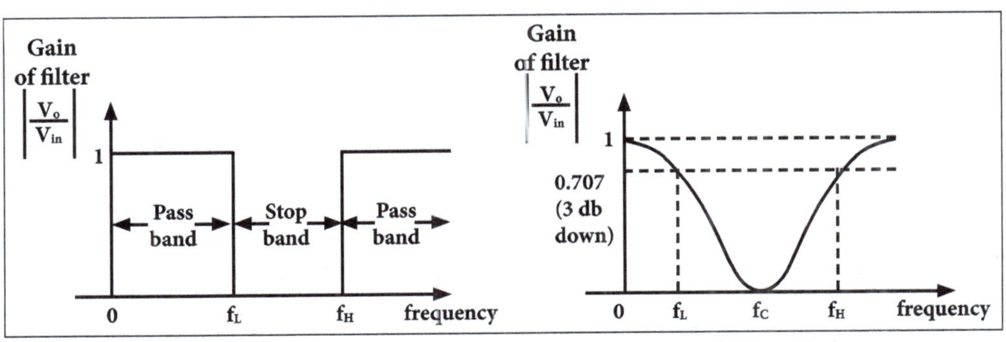

(a) Ideal (b) Practical

Band Reject

At frequency $f = f_c$, the practical characteristics shows a notch and hence, this filter is also called Notch filter. The frequency f_c in both band pass and band elimination filters is called Centre frequency as it is approximately at the centre of the pass band and stop band respectively for the two filters.

All Pass Filters

The all pass filter passes all the frequencies but it produces the phase shift between the input and output. The output and input voltages are equal in amplitude for all the frequencies but with the phase shift between the two.

The frequency up to which the input-output amplitude remains same is decided by the unity gain bandwidth (UCB) of the op-amp used. The characteristics showing phase shift between input and output as shown in the below figure:

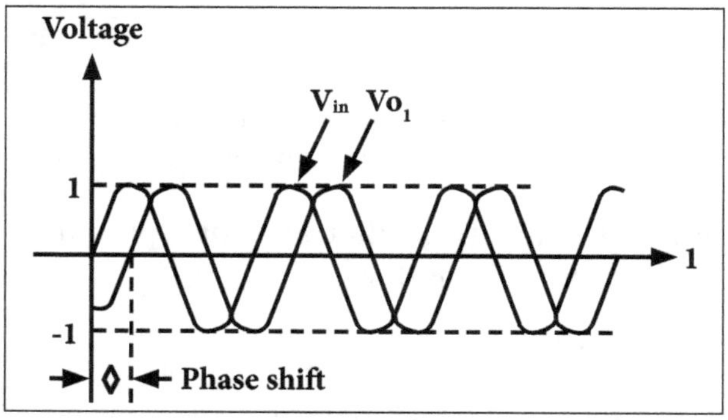

All Pass Filters.

2.2 Analysis of 1st Order LPF and HPF Butterworth Filters

Depending on the type of elements used in their construction, filter may be classified as passive or active. Elements used in passive filters are resistors, capacitors and inductors. Elements used in active filters are transistor or op-amp.

Active filters offer the following advantages over a passive filter:

- No loading problem: Because of its high input resistance and low output resistance of the op-amp, the active filter does not cause loading of the source or load.
- Cost: Active filters are more economical than passive filter. This is because of the variety of cheaper op-amps and the absence of inductors.
- Gain and Frequency adjustment flexibility: Since the op-amp is capable of providing a gain, the input signal is not attenuated as it is in the passive filter.

The most commonly used filters are:

- Low pass Filters.

- High pass Filters.
- Bands pass filters.
- Band rejects filters.
- All pass filters.

The low-pass filter has a gain response with a frequency range from zero frequency (DC) to ω_C. Any input that has a frequency below the cut-off frequency ω_C gets a pass and anything above it gets attenuated or rejected. The gain approaches zero as frequency increases to infinity.

The input signal of the filter shown here has equal amplitudes at frequencies, ω_1 and ω_2. After passing through the low-pass filter, the output amplitude at ω1 is unaffected because it is below the cutoff frequency ω_C. However, at ω_2, the signal amplitude is significantly decreased because it's above ω_C.

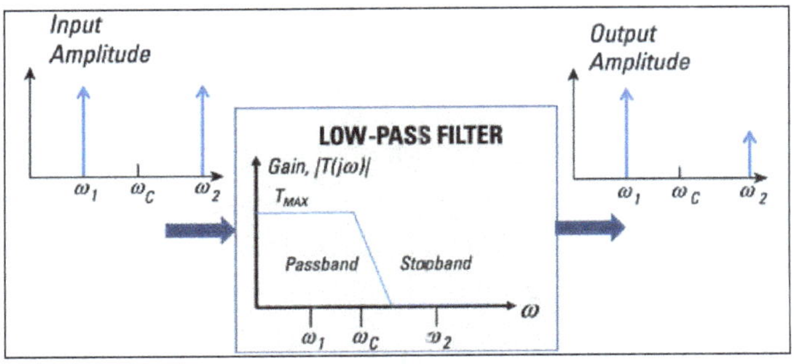

Low Pass Filters.

The high-pass filter has a gain response with a frequency range from ω_C to infinity. Any input having a frequency below the cut-off frequency ω_C gets attenuated or rejected. Anything above ω_C passes through unaffected.

The input signal of the filter shown here has equal amplitude at frequencies ω_1 and ω_2. After passing through the high-pass filter, the output amplitude at ω1 is significantly decreased because it is below ω_C and at ω_2, the signal amplitude passes through unaffected because it is above ω_C.

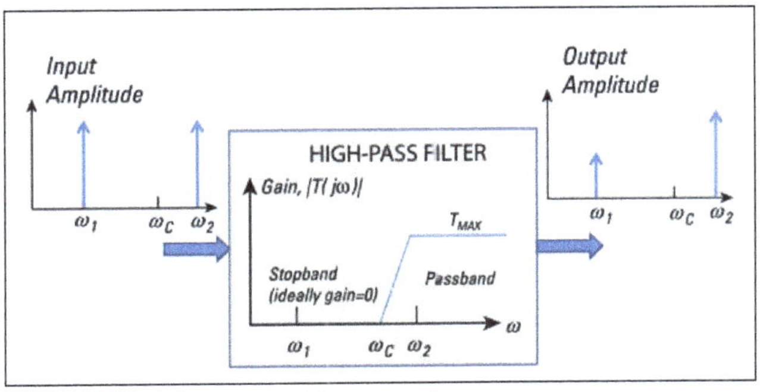

High-Pass Filter.

Active First Order Low Pass Filter

The most common and easily understood active filter is the active low pass filter. Its principle of operation and frequency response is exactly the same as those for the passive filter the only difference this time is that it uses an op-amp for amplification and gain control. The simplest form of a low pass active filter is to connect an inverting or non-inverting amplifier.

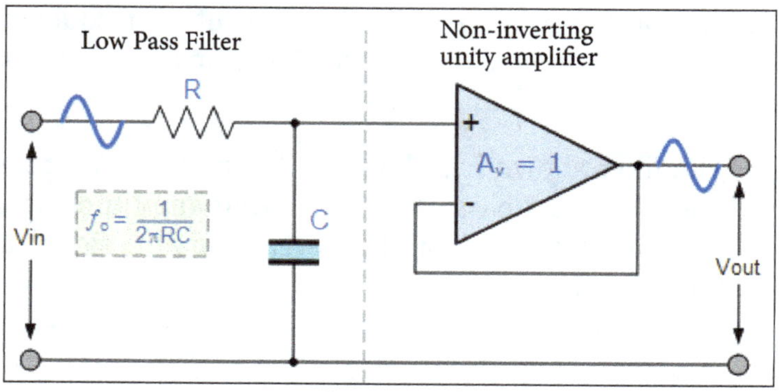

First Order Low Pass Filter.

This first-order low pass active filter simply consists of a passive RC filter stage that provides a low frequency path to the input of non-inverting operational amplifier. The amplifier configuration is carried out as a voltage-follower giving it DC gain of one, $A_v = +1$ or unity gain as opposed to passive RC filter that has a DC gain of less than unity.

The advantage of this type of configuration is that the op-amps of high input impedance prevents excessive loading on the filter's output and its low output impedance prevents filters cut-off frequency point from affected by the changes in the impedance of load.

By the time, this configuration provides a good stability to the filter its main disadvantage is that it provides no voltage gain above one. Even though the voltage gain is unity, the power gain is very much high as its output impedance is much lower than the input impedance. When a voltage gain greater than one is required we may use the following filter circuit.

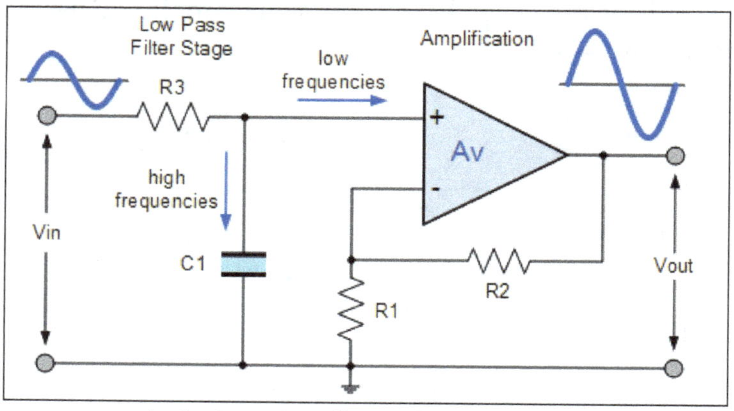

Active Low Pass Filter with Amplification.

The frequency response of this circuit will be same as that of the passive RC filter, except that the amplitude of the output is increased by pass band gain, A_F of amplifier. In case of non-inverting amplifier circuit, the magnitude of voltage gain for the filter is given by a function of the feedback resistor (R_2) divided by its corresponding input resistor (R_1) value and this is given as,

$$\text{DC gain} = \left(1 + \frac{R_2}{R_1}\right)$$

Hence, the gain of an active low pass filter as the function of frequency will be,

Gain of first-order low pass filter,

$$\text{Voltage Gain}(Av) = \frac{V_{out}}{V_{in}} = \frac{A_F}{\sqrt{1 + \left(\frac{f}{fc}\right)^2}}$$

Where,

- A_F = the pass band gain of the filter, $(1 + R_2/R_1)$.
- f = frequency of the input signal in Hertz, (Hz).
- f_c = cut-off frequency in Hertz, (Hz).

Therefore, the operation of a low pass active filter is verified from the frequency gain equation above which is given by,

- At very low frequencies, $f < fc \;\; \frac{V_{out}}{V_{in}} \cong A_F$.
- at the cut-off frequency, $f = fc \;\; \frac{V_{out}}{V_{in}} = \frac{A_F}{\sqrt{2}} = 0.707\,A_F$.
- At very high frequencies, $f > fc \;\; \frac{V_{out}}{V_{in}} < A_F$.

Hence, the active low pass filter consists of a constant gain AF from 0 Hz to the high frequency cut-off point, f_C. At f_C the gain is $0.707A_F$, and after f_C the value decreases at a constant rate as the frequency increases. That is, when the frequency is increased tenfold, then the voltage gain is divided by 10.

In other words, gain decreases 20dB (= 20log 10) each time when the frequency is increased by 10. When dealing with the filter circuits the magnitude of the pass band gain of the circuit is basically expressed in decibels or dB as a function of the voltage gain, and this can be defined as,

Magnitude of the voltage gain in (dB),

$$Av\ (dB) = 20\log_{10}\left(\frac{V_{out}}{V_{in}}\right)$$

$$\therefore -3dB = 20\log_{10}\left(0.707\frac{V_{out}}{V_{in}}\right)$$

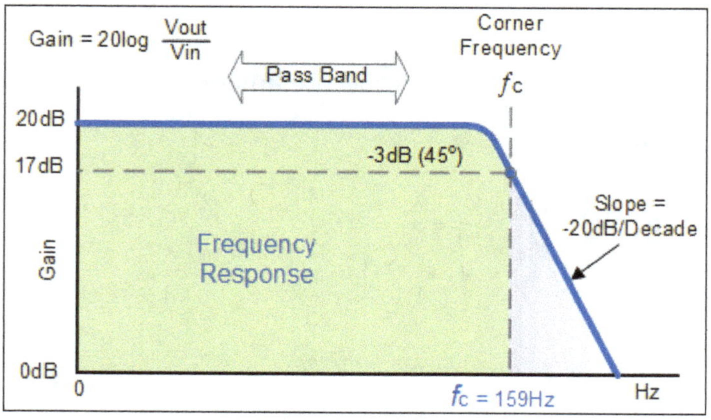

Frequency Response Curve.

When the external impedance connected to the input of circuit changes, even this change will also affect the corner frequency of the filter. One way of avoiding this is by placing the capacitor in parallel with feedback resistor R_2.

This will result in the value of the capacitor, by changing it slightly from being 100nF to 110nF to take account of the 1Ω resistor. The formula used to calculate the cut-off corner frequency is the same as that formula used for the RC passive low pass filter.

$$f_C = \frac{1}{2\pi CR_2} \text{Hertz}$$

Second-Order Low Pass Active Filter

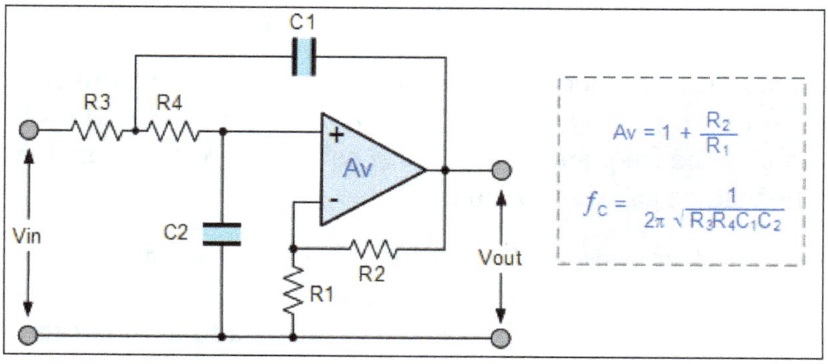

Second-Order Active Low Pass Filter Circuit.

Since with the passive filter, a first-order low pass active filter can be converted into a second-order low pass filter just by using an additional RC network in the input path. The frequency response of second-order low pass filter becomes identically same to that of first-order type, except that stop band roll-off will be twice the first-order filters at 40dB/decade. Thus, the design steps required of the second-order active low pass filter are same.

When cascading together with the filter circuits to form higher-order filters, overall gain of the filter becomes equal to the product of each stage. For instance, the gain of one stage might be 10 and gain of the second stage might be 32 and the gain of a third stage might be 100. Then the overall gain may be 32,000, as shown below:

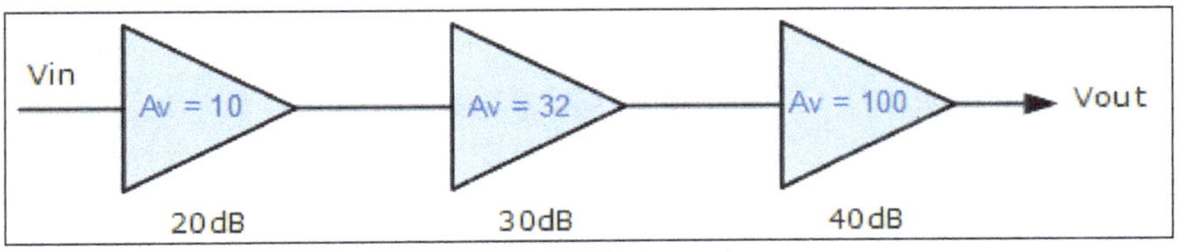

Cascading Voltage Gain.

$$Av = Av_1 \times Av_2 \times Av_3$$

$$Av = 10 \times 32 \times 100 = 32{,}000$$

$$Av\ (dB) = 20\log_{10}(32{,}000)$$

$$Av\ (dB) = 90\,dB$$

$$90\,dB = 20\,dB + 30\,dB + 40\,dB$$

Second-order active filters are important because of the higher-order filters which can be designed using them. By cascading together the first and the second-order filters, filters with an order value, either odd or even up to any value may be constructed.

Problems

1. Let us design an OP-AMP based second order active low pass filter with the cut off frequency 2 kHz.

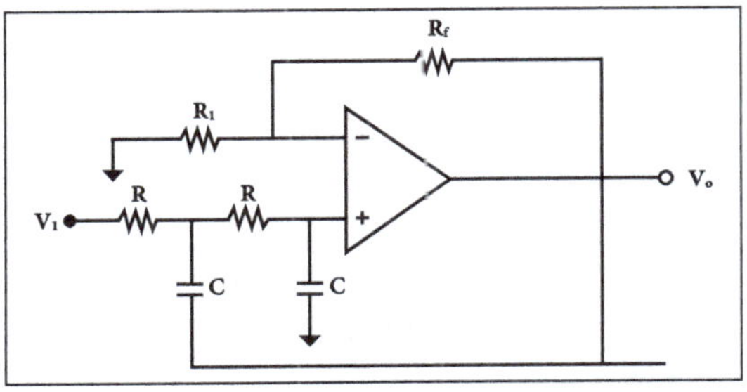

Solution:

Given:

$$N = 2$$

$$f = 2kHz$$

Formula:

$$= \frac{1}{2}\pi RC$$

Let,

$$C = 0.1\ \mu F$$

$$R = 0.8\ kW$$

$$\alpha = 1.414$$

The pass band gain $A = 3 - \alpha$

$$= 3 - 1.414$$

$$= 1.586$$

Thus, the transfer function of second order low pass filter is,

$$H(s) = \frac{A}{S^2 + 1.414S + 1} = \frac{1.586}{S^2 + 1.414S + 1}$$

Here,

$$A = 1 + \frac{R_f}{R_i} = 1.586$$

Hence,

$$R_i = 10\,k\Omega$$

Active First Order High Pass Filters

Basic electrical operation of an active high pass filter is exactly same as that of an equivalent RC passive high pass filter circuit, except at this time the circuit consist of an operational amplifier or an op-amp included within the filter design to provide the amplification and the gain control.

Just like the active low pass filter circuit, the simplest form of an active high pass filter is to connect standard inverting or non-inverting operational amplifier to basic RC high pass passive filter circuit as shown in the figure.

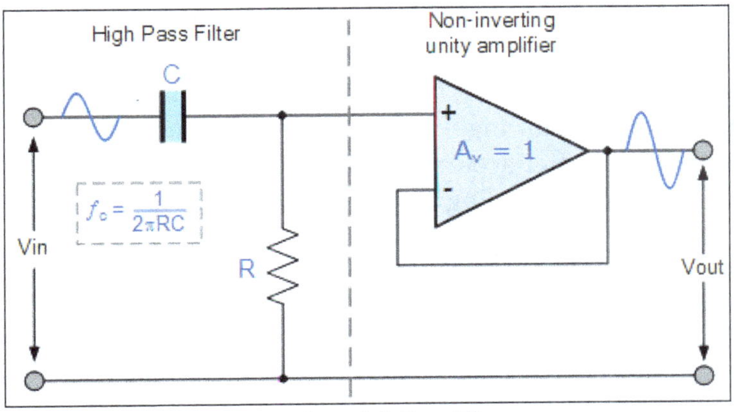

First Order High Pass Filter.

But technically, there is nothing called as an active high pass filter. Unlike the passive high pass filters which have an "infinite" frequency response, the maximum pass band frequency response of an active high pass filter is always limited by the open-loop characteristics or by the bandwidth of an operational amplifier being used, which makes them appear as if they are the band pass filters with a high frequency cut-off found by the selection of an op-amp and gain.

In the operational amplifier, maximum frequency response of an op-amp is limited to gain/bandwidth product or the open loop voltage gain (AV) of the operational amplifier being used giving it a bandwidth limitation, where closed loop response of the op amp intersects with the open loop response.

The commonly available operational amplifier like the uA741 has a "open-loop" DC voltage gain of about 100dB maximum reducing at roll off rate of -20dB/decade as input frequency increases. The gain of the uA741 keeps reducing until it reaches unity gain, or its "transition frequency" (ft) that is about 1MHz. This causes the op-amp to have the frequency response curve similar to that of first-order low pass filter and this is shown in the below figure:

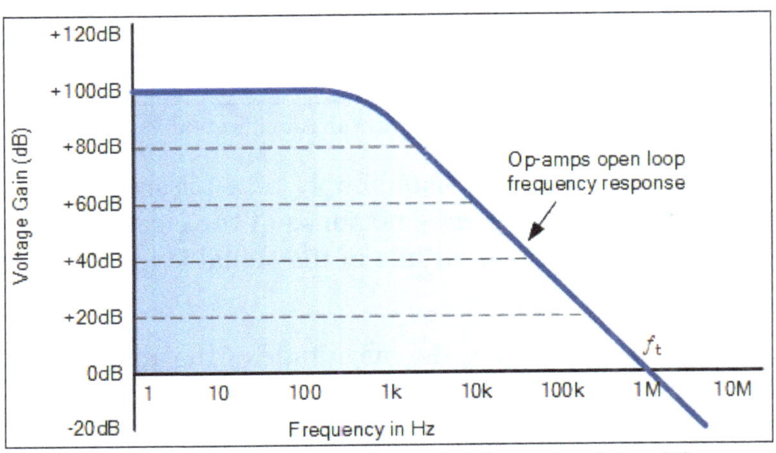

Frequency Response Curve of a Typical Operational Amplifier.

Then the performance of a "high pass filter" at high frequencies is limited by this unity gain crossover frequency which determines the overall bandwidth of open-loop amplifier. The gain-bandwidth product of op-amp starts from around 100kHz for small signal amplifiers up to 1GHz for high-speed digital video amplifiers and op-amp based active filters may achieve better accuracy and performance provided that low tolerance resistors and the capacitors are being used.

Under normal circumstances the maximum pass band needed for a closed loop active high pass or band pass filter is well below than that of the maximum open-loop transition frequency. But, while designing active filter circuits, it is important to choose the correct op-amp for the circuit as loss of high frequency signals will result in signal distortion.

A first-order active high pass filter as its name implies, attenuates low frequencies and passes the high frequency signals. It simply has a passive filter section followed by non-inverting operational amplifier.

The frequency response of the circuit is the same as that of the passive filter, except that the amplitude of the signal is being increased by the gain of the amplifier and for non-inverting amplifier the value of the pass band voltage gain is given as 1 + R2/R1, the same as for the low pass filter circuit.

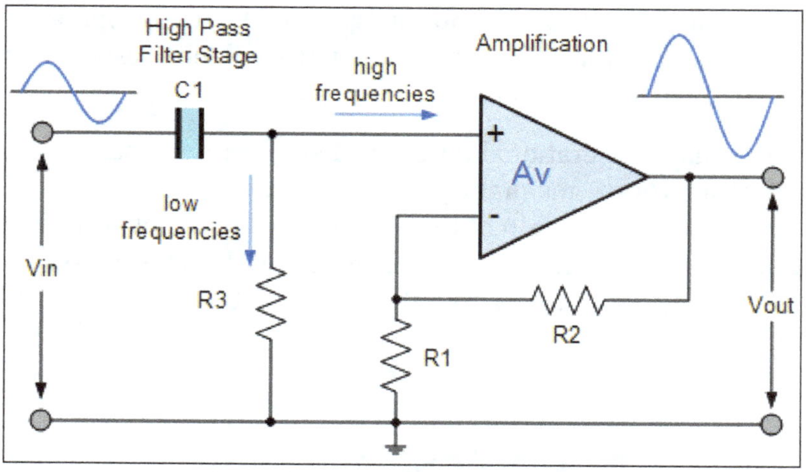

Active High Pass Filter with Amplification.

The first-order high pass filter consists simply of a passive filter followed by a non-inverting amplifier. The frequency response of the circuit is the same as that of passive filter, except that the amplitude of the signal is increased by the gain of amplifier.

For a non-inverting amplifier circuit, the magnitude of the voltage gain for the filter is given by a function of the feedback resistor (R_2) divided by its corresponding input resistor (R_1) value and is given by,

Gain for an active high pass filter,

$$\text{Voltage Gain}(Av)=\frac{V_{out}}{V_{in}}=\frac{A_F\left(\frac{f}{f_c}\right)}{\sqrt{1+\left(\frac{f}{f_c}\right)^2}}$$

Where,

f = Frequency of the Input Signal in Hertz, (Hz)

A_F = Pass band Gain of the filter, $(1+R_2/R_1)$

f_c = Cut-off Frequency in Hertz, (Hz)

Just like the low pass filter, the operation of a high pass active filter can be verified from the frequency gain equation above as,

- At very low frequencies, $f<fc$ $\frac{V_{out}}{V_{in}}<A_F$.
- At the cut-off frequency, $f=fc$ $\frac{V_{out}}{V_{in}}=\frac{A_F}{\sqrt{2}}=0.707\,A_F$.
- At very high frequencies, $f>fc$ $\frac{V_{out}}{V_{in}}\cong A_F$.

Then, the active high pass filter has gain AF that increases from 0Hz to low frequency cut-off point, fC at 20dB/decade as frequency increases. At f_C the gain is 0.707AF, and after f_C all frequencies are pass band frequencies so the filter has a constant gain AF with the highest frequency that is determined by the closed loop bandwidth of op-amp.

While dealing with filter circuits the magnitude of the pass band gain of the circuit is basically expressed in decibels or dB as function of voltage gain, and this is referred to as the,

Magnitude of the voltage gain in (dB),

$$Av(dB)=20\log_{10}\left(\frac{V_{out}}{V_{in}}\right)$$

$$\therefore\ -3dB=20\log_{10}\left(0.707\frac{V_{out}}{V_{in}}\right)$$

In case of a first-order filter, the frequency response curve of the filter is increased by 20dB/decade or 6dB/octave up to the determined cut-off frequency point which always is at -3dB below maximum gain value. Since with the filter circuits, the lower cut-off or corner frequency (f_c) can be determined by using the same formula.

$$f_C = \frac{1}{2\pi RC}\text{Hz}$$

The corresponding phase angle or phase shift of output signal is as same as for the passive RC filter and leads that of the input signal. It is equal to +45° at cut-off frequency fc value and this formula is given as,

$$\text{Phase Shift } \phi = \tan^{-1}\left(\frac{1}{2\pi f RC}\right)$$

Simple first-order active high pass filter may also be made using inverting operational amplifier configuration as well. An example of this circuit design is given with its corresponding frequency response curve. Gain of 40dB is assumed for the circuit.

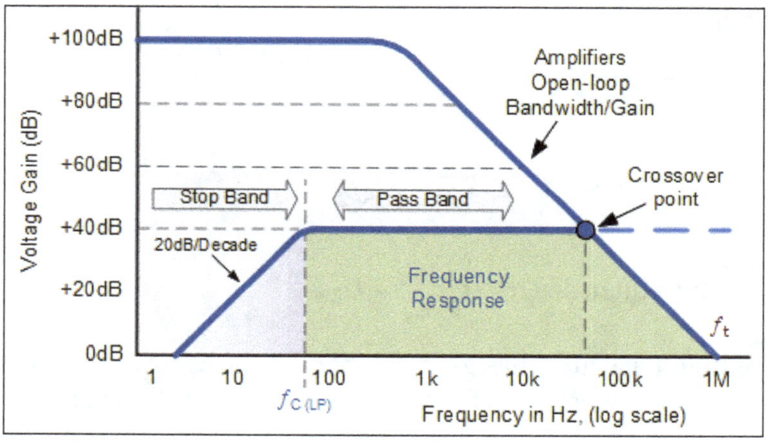

Frequency Response Curve.

Second-Order High Pass Active Filter

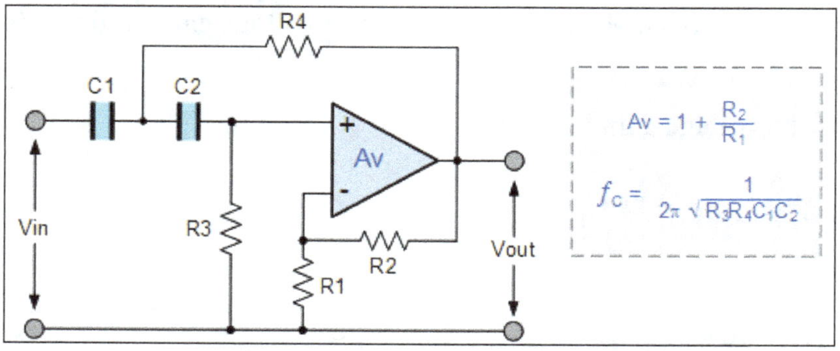

Second-Order Active High Pass Filter Circuit.

Since with passive filter, first-order high pass active filter can be converted into a second-order high pass filter just by using an additional RC network in the input path. The frequency response of second-order high pass filter is same as that of the first-order type except that the stop band roll-off will be twice the first-order filters at 40dB/decade. Therefore, the design steps required of the second-order active high pass filter are the same.

Higher-order high pass active filters, such as third, fourth, fifth, etc are formed simply by cascading together first and second-order filters.

For example, a third order high pass filter is formed by cascading in series first and second order filters, a fourth-order high pass filter by cascading two second-order filters together and so on.

Then the active high pass filter with an even order number will have only second-order filters, while an odd order number will start with a first-order filter at the starting as shown:

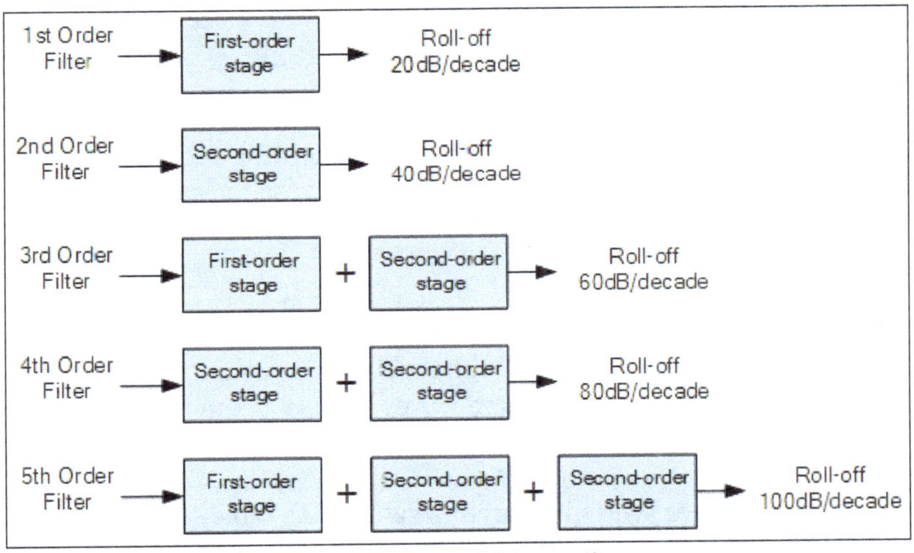

Cascading Active High Pass Filters.

Even though there is no limit to the order of filter that may be formed, as the order of the filter increases so does its size. Also, the accuracy of which declines that is the difference between actual stop band response and the theoretical stop band response increases.

If the frequency determining resistors are equal, $R_1 = R_2 = R_3$ etc, and the frequency finding capacitors are equal, $C_1 = C_2 = C_3$ etc, then the cut-off frequency for any order of filter will be exactly same. However, the overall gain of the higher-order filter is fixed because all the frequency determining components are equal.

2.3 Waveform Generators: Triangular, Saw Tooth and Square Wave

Waveform generators are electronic circuits which can produce sinusoidal, square, triangular, and saw tooth waveforms using oscillators and pulsed circuits.

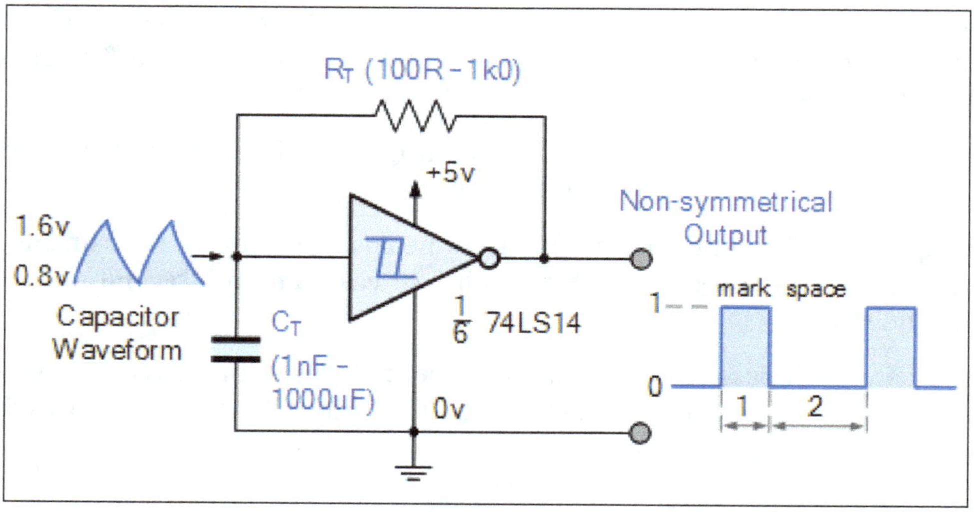

Waveform Generators.

But it is also possible to construct basic Waveform Generator circuits from simple integrated circuits or operational amplifiers connected to a resistor-capacitor (RC) tank circuit or to a quartz crystal to produce the required binary or square wave output waveform at the desired frequency.

This waveform generation would be incomplete without some examples of digital regenerative switching circuits, since it illustrates both the switching action and operation of waveform generators used for generating square waves for use as timing or sequential waveforms.

We know that regenerative switching circuits such as As table Multi vibrators are the most commonly used type of relaxation oscillator as they produce a constant square wave output, making them ideal as a digital Waveform Generator.

As table multi vibrators make excellent oscillators because they switch continuously between their two unstable states at a constant repetition rate thereby producing a continuous square wave output with a 1 : 1 mark-space ratio ("ON" and "OFF" times the same) from its output and we will look at some of the different ways we can construct waveform generators using just standard TTL and CMOS logic circuits along with some additional discrete timing components.

2.3.1 Triangular Waveform Generator

A triangular wave generator can be formed by simply connecting an integrator to a square wave generator. The resultant circuit is shown in the below figure. This circuit requires dual op-amp, two capacitors and at least five resistors.

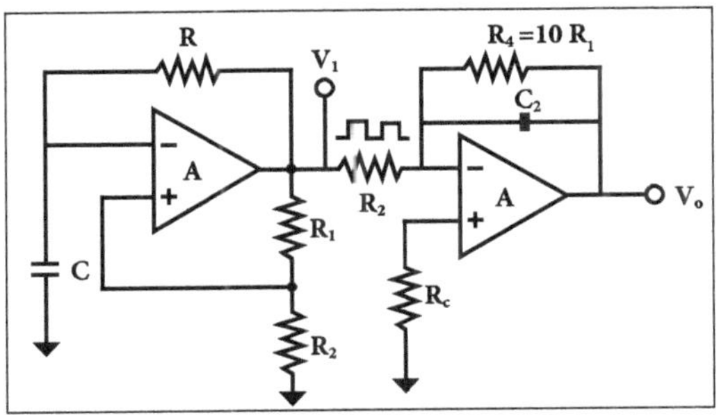

Triangular Wave Generator.

The frequencies of the square wave and triangular wave are the same. For fixed R_1, R_2 and C values, the frequency of the square wave as well as the triangular wave depends on the resistance R.

As R is increased or decreased, the frequency of the triangular wave will decrease or increase respectively. Although the amplitude of the square wave is constant at $\pm V_{sat}$, the amplitude of the triangular wave will decrease as the frequency increases. This is because the reactance of the capacitor C in the feedback circuit decreases at high frequencies.

A resistance R_4 is connected across C_2 to avoid the saturation problem at low frequencies as in the case of practical integrator.

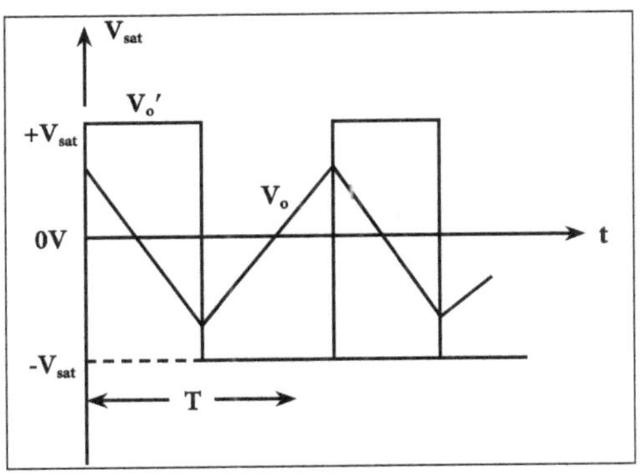

Output Waveform.

Another, triangular wave generator using lesser number of components is shown in the below figure:

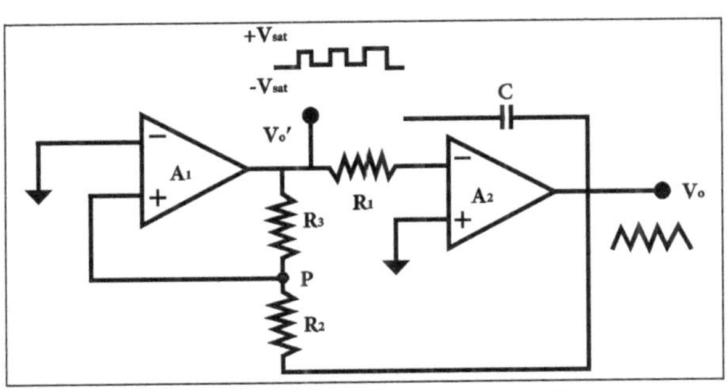

Triangular Wave Generator Using Lesser Components.

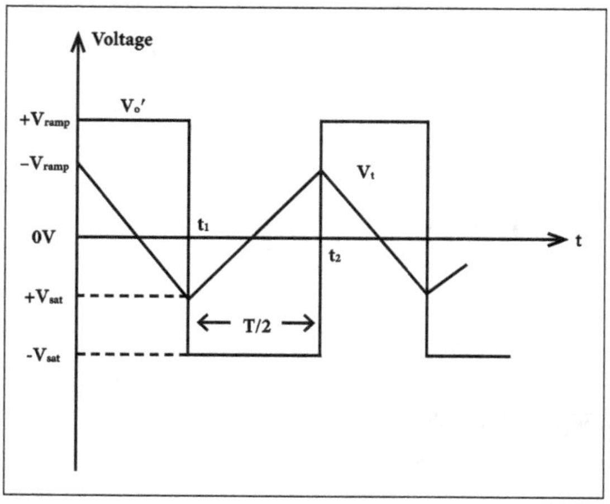

Output Waveforms.

The triangular wave generator consists of a two level comparator and an integrator A_2. The output of comparator A_1 is a square wave and is applied to the (-) input terminal of the integrator A_2 producing a triangular wave. This triangular wave is fed back as the input to the comparator A1 through a voltage divider.

Initially, let us consider that the output of comparator A_1 is $+V_{sat}$. The output of the integrator A_2 will be a negative going ramp as shown in the above figure. Thus, one end of the voltage divider R_2 R_3 is at a voltage +Vsat and the other at the negative going ramp of A_2.

At a time $t = t_1$, when the negative going ramp attains a value of -V, the voltage at point P becomes slightly less than 0V. This switches the output of A_1 from positive saturation to negative saturation level - V_{sat}.

During the time when the output of A1 is at $-V_{sat}$, the output of A_2 increases in the positive direction. And at the instant $t = t_2$, the voltage at point P becomes just above 0V, thereby switching the output of A1 from $-V_{sat}$ to $+V_{sat}$.

The cycle repeats and generates a triangular waveform. It can be seen that the frequency of the square wave and triangular wave will be same. However, the amplitude of the triangular wave depends upon the RC value of the integrator A_2 and the output voltage level of A_1.

The output voltage can be set to desired level by using appropriate zener diodes. The frequency of the triangular waveform can be calculated as follows:

The effective voltage at the point P during the time when the output of A1 is at +Vsat level is given by,

$$-V_{ramp} + \frac{R_2}{R_2 + R_3}\left[+V_{sat} - \left(-V_{ramp}\right)\right] \quad ...(1)$$

At $t = t_1$, the voltage at point P becomes equal to zero. Therefore, from the above equation,

$$-V_{ramp} = -\frac{R_2}{R_3}(+V_{sat}) \qquad ...(2)$$

Similarly, at t = t2 when the output of A1 switches from –Vsat to +Vsat.

$$V_{ramp} = \frac{-R_2}{R_3}(-V_{sat})$$

$$= \frac{R_2}{R_3}V_{sat} \qquad ...(3)$$

Therefore, peak to peak amplitude of the triangular wave is,

$$V_{O(PP)} = +V_{ramp} - (-V_{ramp})$$

$$V_{O(PP)} = 2\frac{R_2}{R_3}V_{sat} \qquad ...(4)$$

The output switches from – Vramp to + Vramp in half the time period T/2. Putting the values in the basic integrator equation,

$$V_o = -\frac{1}{RC}\int V_i\,dt)$$

$$V_o\,Cpp) = \frac{1}{R_1C_1}\int_0^{T/2}(-V_{sat})dt = \frac{V_{sat}}{R_1C_1}\left(\frac{T}{2}\right)$$

$$T = 2R_1C_1\frac{V_o(pp)}{V_{sat}}$$

Putting the value of $V_o(pp)$ from equation (4) we get,

$$T = \frac{4R_1C_1R_2}{R_3}$$

Hence, the frequency of oscillation fo is given by,

$$f_o = \frac{1}{T} = \frac{R_3}{4R_1C_1R_2} \qquad ...(5)$$

2.3.2 Saw Tooth Waveform Generator

The saw tooth wave refers to a waveform with its rise time being many times longer than the corresponding fall time or fall time very longer as compared to the use time. Triangular wave generator can be modified to produce a saw tooth waveform.

The saw tooth wave generator circuit is shown in the below figure (a). The op-amp A_1 functions as a ramp generator and the op-amp A_2 function as a comparator. The input reference signal V_i of value less than zero is connected to the inverting input of the op-amp. Since V_i is negative, the output of op-amp A_1 can only ramp up.

The rate of raise is given by $\frac{V_{oR}}{t} = \frac{V_i}{R_i C}$. The ramp voltage VoR is monitored by the comparator A2.

The output V_{oR} is connected to the non-inverting terminal of op-amp A_2, and a reference voltage set by a potentiometer V_{ref} is connected to the inverting input of comparator. When the capacitor C charges, and when the voltage V_{oR} is below V_{ref}, the output of comparator is negative. Then, the transistors Q_1 and Q_2 do not conduct. The diodes D_1 and D_2 protect the transistors from excessive reverse bias voltages.

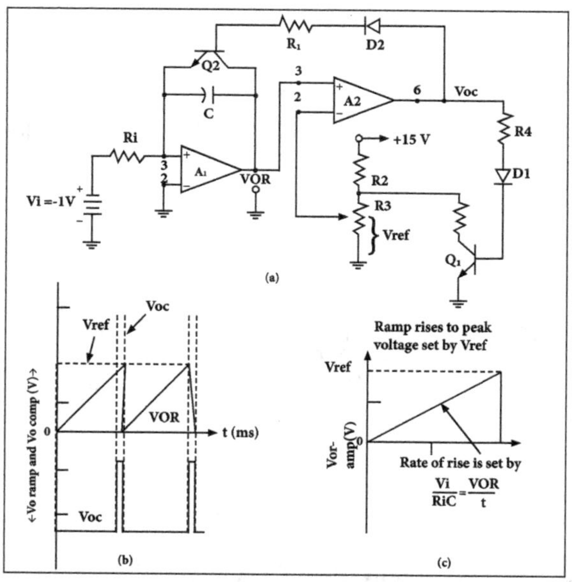

Saw tooth Wave Generator (a) Circuit Diagram (b) Saw tooth Wave and Comparator Outputs (c) Saw tooth Wave Design.

When V_{oR} rises and just exceeds V_{ref}, the output V_{oc} of comparator goes to positive saturation. This action forward biases the transistor Q_2 into saturation. The saturated transistor then acts as a switch across the capacitor C, which discharges quickly making the output V_{oc} come down essentially to 0 V.

The positive saturation V_{oc} of A_2 also makes the transistor Q_1 ON, and the V_{ref} or negative input of op-amp A_2 drops to 0 V. As the capacitor C discharges rapidly making V_{oR}

zero volts, it drops below V_{ref}. This causes the comparator output to become negatively saturated. This action switches the transistor Q_2 OFF and C begins charging linearly and the cycle repeats.

Frequency of Oscillation

The charging time Tc is obtained from period $T = \frac{V_{ref}}{V_1 / R_i C}$. The frequency is the reciprocal of the time period.

Therefore,

$$f = \frac{1}{T} = \left(\frac{1}{R_i C}\right)\frac{V_1}{V_{ref}}$$

The saw tooth output waveform is shown in the above figure (b). The above figure (c) shows the saw tooth wave designing method. The circuit can also be operated as a current-controlled saw tooth wave generator by applying an external control current sink IS at the (-) input terminal of op-amp A1 of the above figure (a).

Then,

$$f = \frac{I_s}{CV_{ref}}.$$

Here, the op-amp A_1 can preferably be a FET input op-amp with low bias current and good slew rate.

Applications of a saw tooth waveform:

- The applications of a saw tooth waveform are in frequency/tone generation, sampling, thyristor switching, modulation, etc.
- The magnetic field produced by the deflection yoke drags the electron beam on the wave's ramp, creating a scan line.
- The ramp portion must appear as a straight line; otherwise, it indicates the magnetic field, which is produced by the deflection yoke, as not linear. This results in non-linearity and the television image is squished. Hence, on that side of the picture, the electron beam spends more time.
- The saw tooth waveform is most common waveform used to create sounds with subtractive virtual and analog music synthesizers. Therefore, it is used in music.
- The saw tooth is the form of horizontal and vertical deflection signals that are used to generate a raster on monitor screens or CRT based television.
- The stability of electronic components improves, and hence, there is no need to adjust the picture's horizontal or vertical linearity.

- With much lower frequency, the vertical deflection operates in similar way as the horizontal deflection system.
- The magnetic field suddenly gets collapsed on the wave's cliff, which causes the resting position of its electron beam as quickly as possible.
- The positive voltage causes a deflection in one direction; the negative voltage causes the deflection in other and the center- mounted deflection uses the screen area to depict a trace.

2.3.3 Square Waveform Generator

Square wave generators are constructed using the Schmitt trigger inverters such as TTL. It is the easy way to make basic as table waveform generator.

While producing clock or timing signals, the as table multi vibrator produces the square wave generator waveform that switches between the HIGH and the LOW.

As we know that the output of a Schmitt inverter is the opposite or inverse to that of its input. By applying hysteresis this can change its state at different voltage levels. It uses Schmitt trigger action which may change the state between the upper and the lower threshold level as the input voltage signal, increases and decreases about input terminal. Here upper threshold level sets the output and the lower threshold level resets output.

This simple square wave generator circuit has a single TTL 74LS14 Schmitt inverter logic gate with a capacitor connected in between its input terminal and ground. The positive feedback required for the circuit to oscillate that is being provided by the feedback resistor.

Assume that the charge across the capacitor plates is below Schmitt's lower threshold level. Therefore makes the input to the inverter at a logic ZERO level resulting in a logic ONE output level.

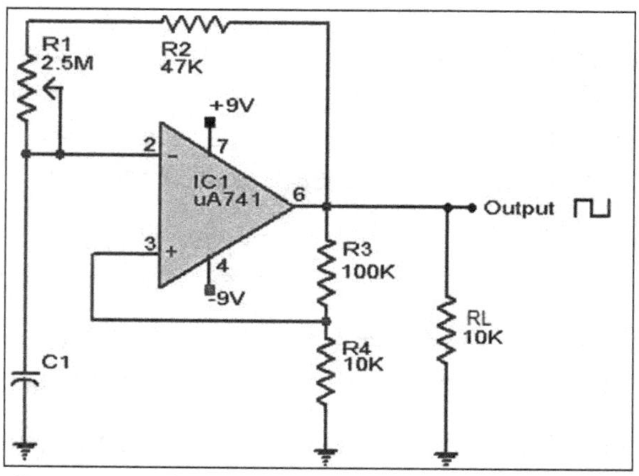

Square Waveform Generator Circuit Using Op-Amp.

Working of Square Wave Generator

The resistor is connected to the logic one level output while the other side of the resistor is connected to the capacitor that is at logic zero level.

Now the capacitor will start charging up in a positive direction through a resistor at a rate obtained by the RC time constant of the combination.

But the charge across the capacitor reaches the upper threshold level of the Schmitt trigger. And the output from Schmitt inverter undergoes a rapid change from a logic level one to logic level zero state. The current flowing through resistor also changes its direction.

The integral of a constant, let us say, 'C' will be the multiple of t, where t is the time across which the integration take place, such that a positive constant will give a positive ramp and a negative constant will integrate to a negative ramp.

Now this change causes the capacitor that was originally charged through the resistor which begins to discharge itself back through the same resistor, until the charge across the capacitor plates reaches the lower threshold level. The inverter output switches states again with the cycle repeating itself over and over again as far as the supply voltage being present.

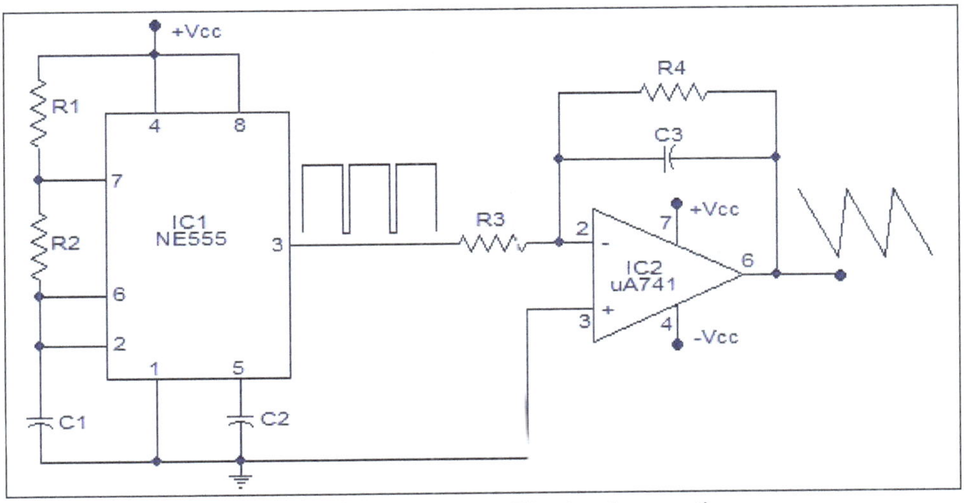

Square Waveform Generator Circuit using 555 timers.

The capacitor is constantly charging and discharging itself during each of the cycle between the inputs of the upper and the lower threshold levels of the Schmitt inverter and producing a logic level ONE or a logic level ZERO at inverter output. The waveform produced is a duty cycle of about one third is as the mark-to-space ratio between HIGH and LOW is 1 to 2 respectively due to the input gate characteristics of the TTL inverter. Hence, the waveform produced is not symmetrical.

2.4 IC555 Timer, Functional Diagram, Mono Stable as Table Operations and Applications

IC Pin Configuration

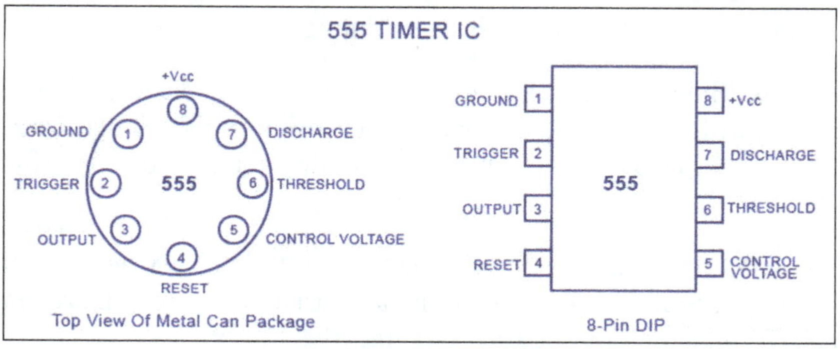

555 Timer IC Pin Configurations.

The 555 Timer IC is available as an 8-pin metal can, an 8-pin mini DIP (dual-in-package) or a 14-pin DIP. The pin configuration is shown in the above figures.

This IC consists of 23 transistors, 2 diodes and 16 resistors. The use of each pin in the IC is explained below. The pin numbers used below refers to 8-pin DIP and 8-pin metal can packages. These pins are explained in detail.

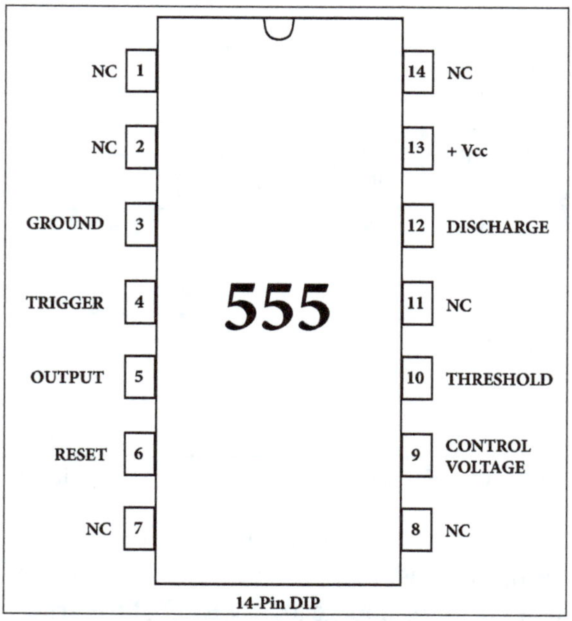

IC Pin Configuration.

Pin 1: Grounded Terminal

All the voltages are measured with respect to the ground terminal.

Pin 2: Trigger Terminal

The trigger pin is used to feed the trigger input then the 555 IC is set up as mono stable multi vibrator. This pin is an inverting input of a comparator and is responsible for the transition of the flip-flop from set to reset.

The output of timer depends on the amplitude of the external trigger pulse applied to this pin. Negative pulse with dc level greater than $V_{cc}/3$ is applied to this terminal. In negative edge, as trigger passes through $V_{cc}/3$, the output of the lower comparator becomes high and complimentary of Q becomes zero. Therefore, the 555 IC output gets a high voltage and thus a quasi-stable state.

Pin 3: Output Terminal

Output of timer is available at this pin. A load can be connected to the output terminal in two ways. One way is to connect between the output pin (pin 3) and ground pin (pin 1) or between the pin 3 and supply pin (pin 8). The load connected between the output and ground supply pin is called the normally on load and that connected between the output and ground pin is called normally off load.

Pin 4: Reset Terminal

Whenever the timer IC is to be disabled or reset, a negative pulse is applied to pin 4 and thus is named as reset terminal. The output is reset irrespective of input condition. When this pin is not to be used for the reset purpose, it should be connected to + V_{CC} to avoid any possibility of false triggering.

Pin 5: Control Voltage Terminal

The threshold and trigger levels are controlled using this pin. Pulse width of output waveform is determined by connecting a POT or bringing in an external voltage to this pin. External voltage applied to this pin can also be used to modulate the output waveform.

Thus, the amount of voltage applied in this terminal will decide when the comparator is to be switched and thus changes pulse width of the output. When this pin is not used, it must be bypassed to ground through a 0.01 micro Farad to avoid any noise problem.

Pin 6: Threshold Terminal

This is the non-inverting input terminal of comparator 1, which compares the voltage applied to terminal with a reference voltage of 2/3 VCC. The amplitude of the voltage applied to this terminal is responsible for the set state of flip-flop. When voltage applied in this terminal is greater than 2/3VCC, the upper comparator switches to +Vsat and output gets reset.

Pin 7: Discharge Terminal

This pin is connected internally to collector of transistor and mostly a capacitor is connected between this terminal and ground. It is called the discharge terminal because when the transistor saturates, capacitor discharges through transistor. When transistor is cut-off, the capacitor charges at a rate determined by external resistor and capacitor.

Pin 8: Supply Terminal

A supply voltage of + 5 V to + 18 V is applied to this terminal with respect to the ground (pin 1).

Working of 555 Timers

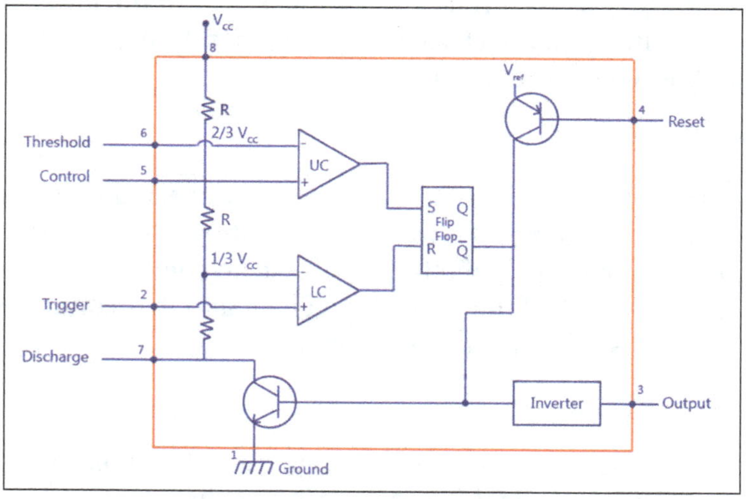

555 IC timer block diagram.

The block diagram of a 555 timer is shown in the above figure. A 555 timer has two comparators, which are basically 2 op-amps), an R-S flip-flop, two transistors and a resistive network:

- Resistive network consists of three equal resistors and acts as a voltage divider.
- Comparator 1 compares threshold voltage with a reference voltage + 2/3 V_{CC} volts.
- Comparator 2 compares the trigger voltage with a reference voltage + 1/3 V_{CC} volts.

Output of both the comparators is supplied to the flip-flop. Flip-flop assumes its state according to the output of the two comparators. One of the two transistors is a discharge transistor of which collector is connected to pin 7. This transistor saturates or cuts-off according to the output state of the flip-flop. The saturated transistor provides a discharge path to a capacitor connected externally. Base of another transistor is connected to a reset terminal. A pulse applied to this terminal resets the whole timer irrespective of any input.

The 555 generally operates in 3 modes:

- As table mode.
- Mono-stable mode.
- Bi-stable mode.

As Table Mode

This means there will be no stable level at the output. So the output will be swinging between high and low. This character of unstable output is used as clock or square wave output for many applications.

Mono-Stable Mode

This configuration consists of one stable and one unstable state. The stable state can be chosen either high or low by the user. If the stable output is set at high (1), the output of the timer is high (1). At the application of an interrupt, the timer output turns low (0). Since the low state is unstable it goes to high (1) automatically after the interrupt passes. Similar is the case for a low stable mono stable mode.

Bi-Stable Mode

In bi-stable mode, both the output states are stable. At each interrupt, the output changes from low (0) to high (1) and vice versa and stays there. For example, if we have a high (1) output, it will go low (0) once it receives an interrupt and stay low (0) till the next interrupt changes the status.

2.4.1 Mono Stable and as Table Operations

Mono Stable Operations

The IC 555 timer can be operated as a mono stable multi vibrator by connecting an external resistor and a capacitor as shown in the below figure:

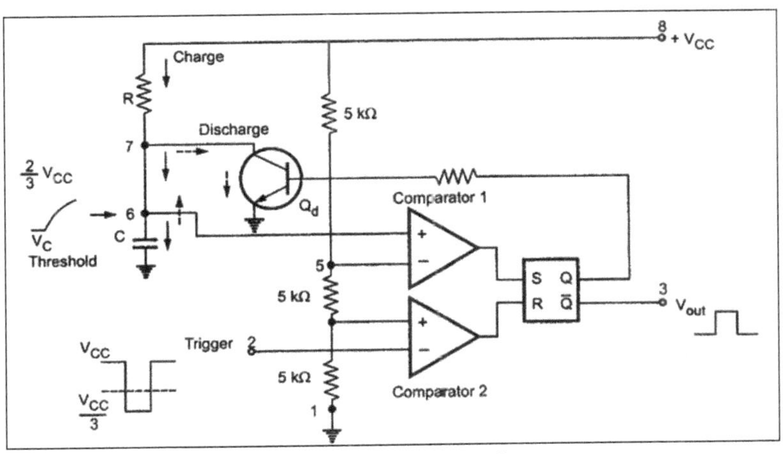

Mono stable operation of 555.

The circuit has only one stable state. When trigger is applied, it produces a pulse at the output and returns back to its stable state. The duration of the pulse depends on the values of R and C. As it has only one stable state, it is called one shot multi vibrator.

Operation

The flip-flop is initially set i.e. Q is high. This drives the transistor Q_d in saturation. The capacitor discharges completely and voltage across it is nearly zero. The output at pin 3 is low.

When a trigger input, a low going pulse is applied, then circuit state remains unchanged till trigger voltage is greater than 1/3 V.

When it becomes less than 1/3 VCC, then comparator 2 output goes high. This resets the flip-flop so Q goes low and (7) goes high. Low Q makes the transistor Qd off. Hence, capacitor starts charging through resistance R, as shown by dark arrows in the above figure.

The voltage across capacitor increases exponentially. This voltage is nothing but the threshold voltage at pin 6. When this voltage becomes more than 2/3 V_{CC}, then comparator 1 output goes high. This sets the flip-flop i.e. Q becomes high and $\overline{Q}$ low. This high Q drives the transistor Qd in saturation. Thus, capacitor C quickly discharges through Qd as shown by dotted arrows in the below figure.

So, it can be noted that Vout, at pin 3 is low at start, when trigger is less than 1/3 V_{CC} it becomes high and when threshold is greater than 2/3 V_{CC} again becomes low, till next trigger pulse occurs. So a rectangular wave is produced at the output. The pulse width of this rectangular pulse is controlled by the charging time of capacitor. This depends on the time constant RC. Thus RC controls the pulse width. The waveforms are shown in the below figure:

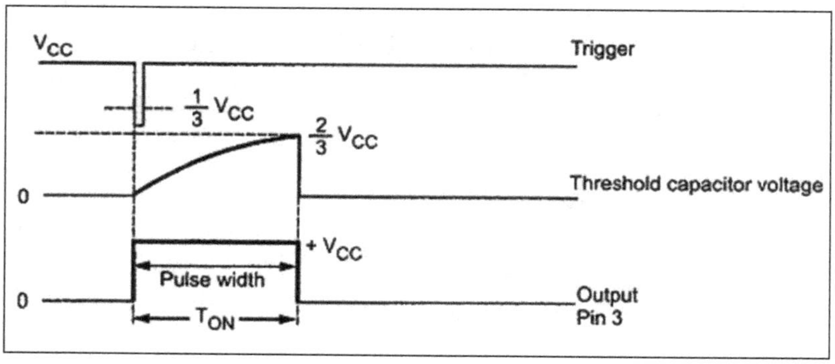

Waveforms of mono stable operation.

Derivation of Pulse Width

The voltage across capacitor increases exponentially and is given by,

$$V_C = V\left(1 - e^{-t/CR}\right)$$

If,

$$V_C = 2/3\ V_{CC}$$

Then,

$$\frac{2}{3}V_{CC}=V_{CC}\left(1-e^{-t/CR}\right)$$

$$\frac{2}{3}-1=-e^{-t/CR}$$

$$\frac{1}{3}=-e^{-t/CR}$$

$$\therefore\ -\frac{t}{CR}=-1.0986$$

$$\therefore\ t=+1.0986\ CR$$

$$\therefore\ t\approx 1.1CR$$

Where C in farads, R in ohms, t in seconds

Thus, we can say that voltage across capacitor will reach 2/3 V_{CC} in approximately 1.1 times, time constant i.e. 1.1 RC.

Thus, the pulse width denoted as W is given by,

$$W=1.1\ RC$$

Schematic Diagram

Generally, a schematic diagram of the IC 555 circuits is shown which does not include comparators, flip-flop etc. It only shows the external components to be connected to the 8 pins of IC 555. Thus, the schematic diagram of IC 555 as a mono stable multi vibrator is shown in the below figure:

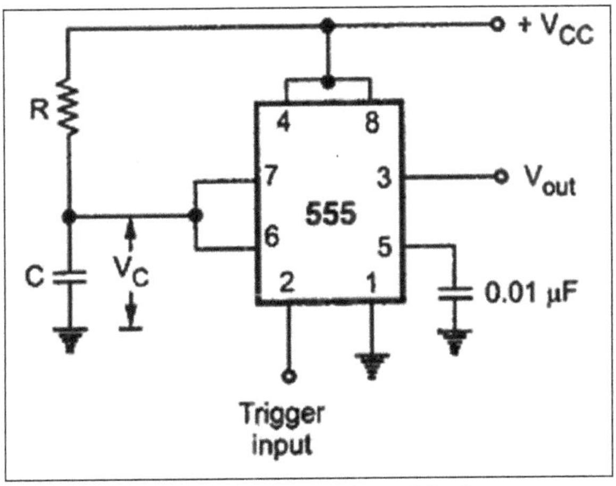

555 timers as mono stable multi-vibrator.

The external components R and C are shown. To avoid accidental reset, pin 4 is connected to pin 8 which is supply +VCC. To have the noise filtering of control voltage, the pin 5 is grounded through a small capacitor of 0.01 μF.

As Table Operations

As table mode of the 555 IC is achieved by connecting resistors R1, R2 and the capacitor C1 to the timer IC as shown in the below figure. R1 is connected between pin 4 and pin 7. While R_2 is connected between pin 7 and pin 6 with the pin 2 as a common node.

Connection of the capacitor C1 is between pin 1 and pin 6. The output produced by the as table circuit is not stable in both the HIGH and the LOW states and it switches continuously from one state to the other. The timing (time period) during which the output is either high or low is determined by the resistors R_1, R_2 and capacitor C_1.

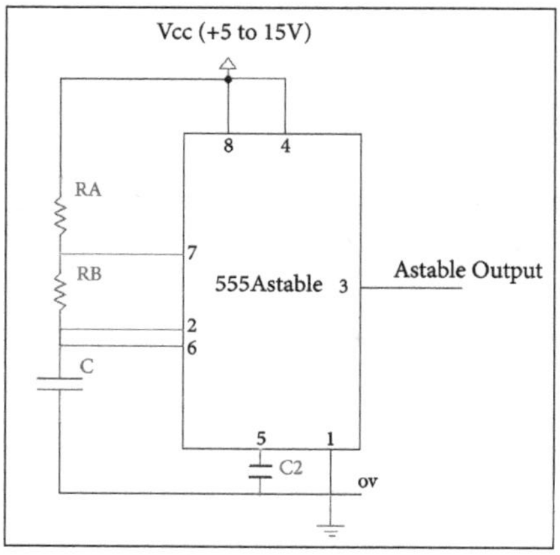

As table Multi-vibrator.

An as table 555 timer circuit usually contains an SR flip flop, a NPN transistor, two comparators and few discrete components like resistors and capacitors for its operation. Initially, when power supply is switched ON, consider that the SR flip flop is cleared with Q = 0.

At this stage, discharging transistor will remain OFF since Q (Q = 0) is connected to the base of the transistor. Capacitor, C will now charge through R_A and R_B to V_{cc}. Hence we can say that, the time constant required for charging is equal to $(R_A + R_B)$ C. As the threshold voltage (capacitor voltage) becomes greater than 2/3 V_{cc}, comparator 1 will be HIGH. The comparator output is HIGH because the voltage at the non-inverting terminal is greater than inverting terminal. Now S = 1, R = 0, Q = 1 and Vout = 0 for the flip flop.

This change in Q will turn the discharging transistor to ON condition. Capacitor C will now turn over to discharging mode. The discharging takes place through the resistor R_B. Hence the discharge time constant will be R_BC. When the capacitor voltage becomes lower than 1/3 V_{cc}, comparator 2 will be HIGH. As a result, S = 0, R = 1, Q = 0 and V_{out} = 1. Discharging transistor will now turn ON and the next cycle of operation will begin.

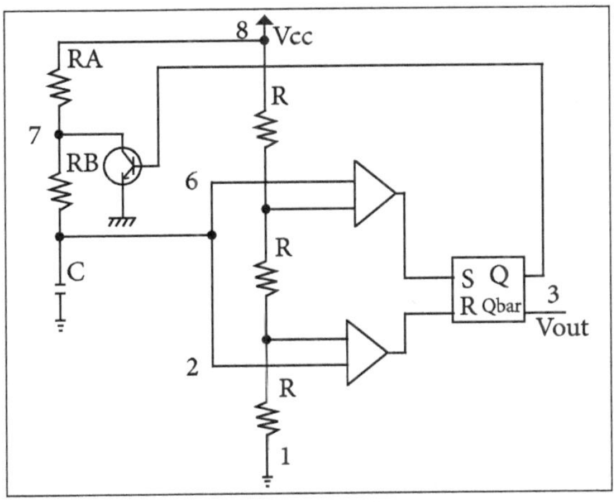

Typical circuit.

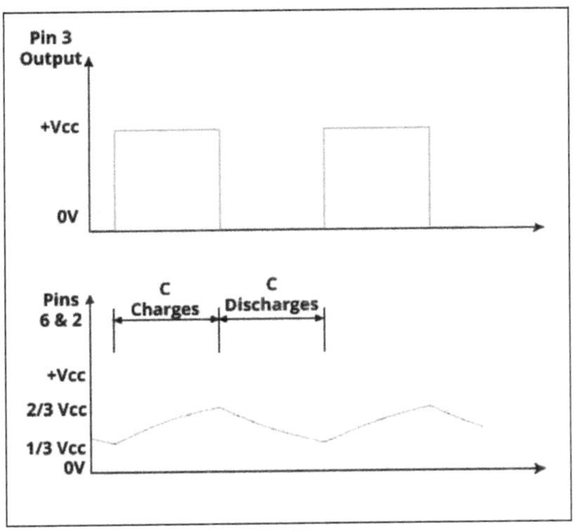

Waveform.

For the 555 IC to operate as an as table multi-vibrator it is necessary to continuously re-trigger the 555 IC after each and every cycle. This re-triggering is achieved by connecting the pin 2 and the pin 6 together. This as table wave oscillator is useful in generating signals for digital circuitry, lamp flashers, sound generators and various other applications.

As Table Multi-Vibrator Charge and Discharge Times

Let t_1 and t_2 represent the charge and discharge times of as table multi-vibrator respectively.

$$t_1 = 0.693(R_A + R_B)$$

$$t_2 = 0.693 R_B C$$

Were R_A and R_B are in ohms and C is in Farad.

As Table Multi-vibrator Cycle Time

Let T be the cycle time of the as table multi-vibrator.

$$T = t_1 + t_2 = 0.693(R_A + 2R_B)C$$

As Table Multi-Vibrator Duty Cycle Time

The duty cycle is defined as ratio of charge time constant (t_1) to the total cycle time (T) and for the as table oscillator it is found to be,

$$\text{Duty Cycle, } D = t_1 / T$$

$$= (R_A + R_B)/(R_A + 2R_B)C$$

Advantages and Disadvantages

The simple operation and less number of external components make the as table multi-vibrator using the 555 IC a popular one. The perfect square wave output from the circuit makes it good for many applications. Some advantages are easy to design, low cost and there is no external triggering. The disadvantage is that we cannot achieve a duty cycle that is less than or equal to 50% using this circuit. This problem can be solved by inverting the output or by using a diode parallel to RB.

2.4.2 Applications

- The mono stable multi-vibrators are used as delay circuits, timers, gated circuits etc.
- One-shot multi-vibrators of both the re-trigger able and non-re trigger able variety find wide application in industry for the siren actuation and machine sequencing, where an intermittent input signal produces an output signal of a set time.
- Mono stable multi-vibrators are used in a number of applications and could be found wherever a square wave or timed interval is necessary for success of a system. For example, the mono stable multi-vibrators were once used in the analog systems to control an output signal's frequency, synchronize line and frame rate of television broadcasts and even moderate tunes of different octaves with electronic organs.
- Additionally, before the integrated circuits invention, mono stable multi-vibrators were connected together in a series to divide frequencies.

2.5 IC565 PLL, Block Schematic, Description of Individual Blocks and Applications

The concept of Phase Locked Loops (PLL) was first emerged in the early 1930's. As the technology was not developed like how it is now, the cost factor for the development of this technology was very high. Because there was a continuous advancement in the field of integrated circuits, PLL has now become one of the main building blocks in the electronics technology. At present, the PLL is available as a single IC in SE/NE560 series (560, 561, 562, 564, 565 and 567) to even reduce the buying cost, the discrete IC's are used to construct a PLL.

The Phase Locked Loops (PLL) is an important building block of linear systems. This technique is used in satellite communication systems air borne navigational systems, FM communication systems, computers etc. PLL's are available as inexpensive monolithic IC's.

Basic Principles

The basic block schematic of the PLL is shown in the below figure.

This feedback system consists of:

- Phase detector / comparator.
- A low pass filter.
- An error amplifier.
- A voltage controlled oscillator (VCO).

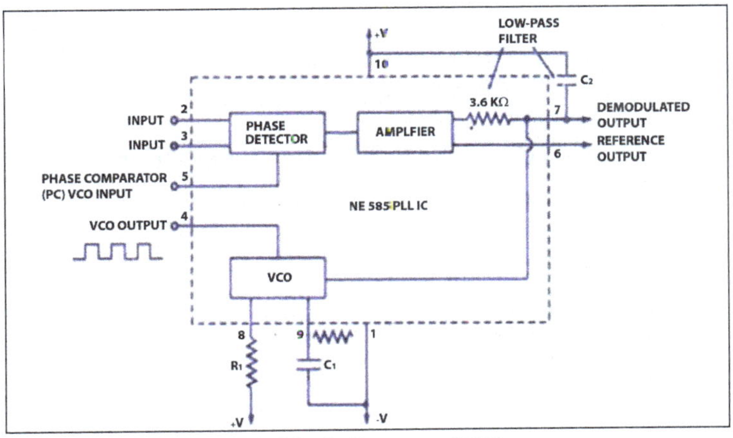

Basic block diagram of PLL.

The VCO is a free running multi-vibrator and operates at a set frequency fo called free running frequency. This frequency is determined by an external timing capacitor and an external resistor.

It can also be shifted to either side by applying a dc control voltage V_C to an appropriate terminal of IC. Its frequency deviation is directly proportional to the control voltage and hence it is termed as voltage controlled oscillator or in short VCO.

If an input signal V_s of frequency f_s is applied to the PLL, phase detector compares the phase and frequency of the incoming signal to that of output V_o of the VCO. If the two signal differ in frequency and/or phase, an error voltage V_e is generated.

The phase detector is basically a multiplier and produces the sum (f_s+f_o) and difference $(f_s - f_o)$ components at its output.

The high frequency component (f_s+f_o) is removed by low pass filter and difference frequency component is amplified and then applied as the control voltage V_C to VCO. The signal VC shifts the VCO frequency in the direction to reduce the frequency difference between f_s & f_o.

Once this action starts, we say that the signal is in the capture range. The VCO continues to change frequency till its output frequency exactly same as the input signal frequency.

The current is said to be locked. Once locked, PLL tracks the frequency changes of the output signal. Thus, a PLL goes through three stages:

- Free running.
- Capture.
- Locked or tracking.

The below figure shows the capture transient. As capture starts, small sine wave appears. This is due to the difference frequency between VCO and input signal. The dc component of the beat drives the VCO towards the locks.

Each successive cycle causes the VCO frequency to move closer to the input signal frequency. The difference in frequency becomes smaller and a larger dc component is passed by the filter, shifting the VCO frequency further.

The process continues until VCO locks ON to the signal and the difference frequency is dc. The LPF controls the capture range. The tracking range is always larger than the capture range.

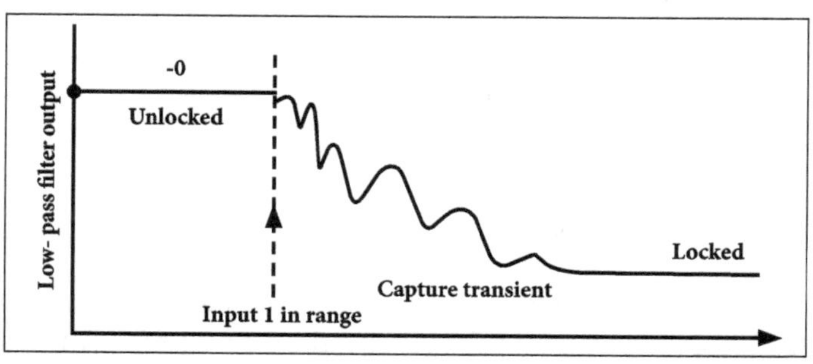

Capture Transient.

The important electrical characteristics of the 565 PLL are:

- Operating frequency range: 0.001Hz to 500 Khz.
- Operating voltage range: ± 6 to ± 12v.
- Input level required for tracking: 10mv rms min to 3 Vpp max.
- Input impedance: 10 K ohms typically.
- Output sink current: 1mA.
- Output source current: 10 mA.

The center frequency of the PLL is determined by the free running frequency of the VCO, which is given by,

$$f_{OUT} = \frac{1.2}{4R1C1} Hz \qquad ...(1)$$

Where R_1 and C_1 are an external resistor & a capacitor connected to pins 8 & 9.

- The VCO free-running frequency f_{OUT} is adjusted externally with R1 & C1 to be at the center of the input frequency range.
- C_1 can be any value R_1 must have a value between 2 k ohms and 20 K ohms.
- Capacitor C_2 connected between 7&+V.
- The filter capacitor C_2 should be large enough to eliminate variations in the de-modulated output voltage in order to stabilize the VCO frequency.
- The lock range fL & capture range fc of PLL is given by,

$$f_L = \pm \frac{8_{fout}}{V} Hz \qquad ...(2)$$

Where f_{OUT} = free running frequency of VCO (Hz)

$$V = (+V) - (-V) \text{volts}$$

$$f_C = \pm \left[f_L / (2\Pi)(3.6)(10^3) C_2 \right]^{1/2} \qquad ...(3)$$

2.5.1 Description of Individual Blocks

Some of the important definitions in relation to PLL are as follows:

Lock in Range

Once the PLL is locked, it can track frequency changes in incoming signals. The range of frequencies over which the PLL can maintain lock with the incoming signal is called lock-in range or tracking range.

The lock-in range is usually expressed as a percentage of f_o, the VCO frequency.

Capture Range

The range of frequencies over which the PLL can acquire lock with an input signal is called capture range. This parameter is also expressed as percentage of fo pull-in time. Total time taken by the PLL to establish the lock is called pull-in time.

This depends on the initial phase and frequency difference between the two signals as well as on overall loop gain and loop filter characteristics.

IC 565 is available as a 14-pin package and a 10-pin metal can package. The pin and block diagram are shown in the below figure. The output frequency of VCO is $f_o = \frac{0.25}{R_T C_T}$ Hz Where, RT, CT external resistance and capacitance connected to pin-8and 9.

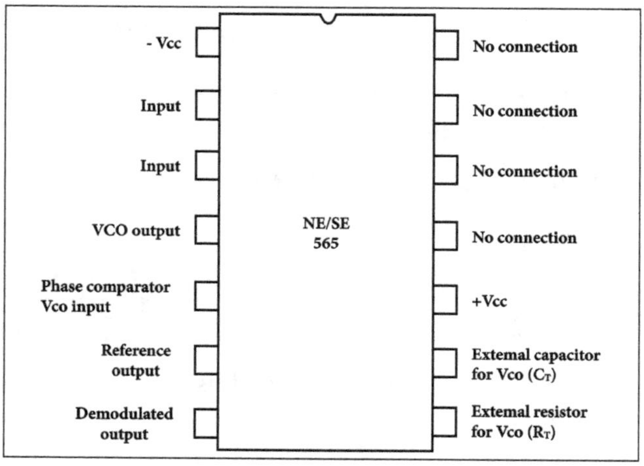

Pin Diagram.

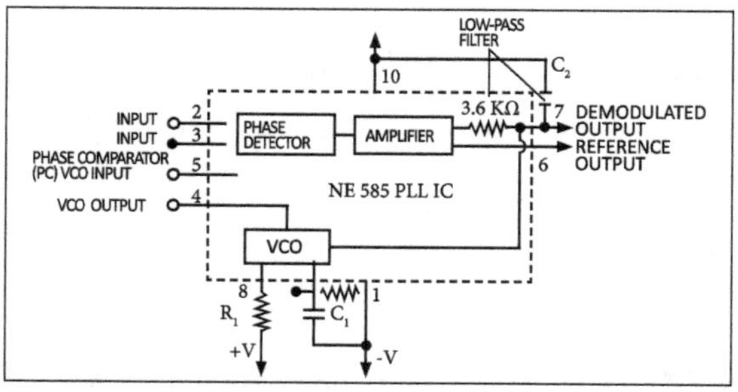

NE/SE 565 PLL block diagram.

$$K\phi = \frac{0.7-(0.7)}{\pi} = \frac{1.4}{\pi}$$

A value between 2KΩ and 20KΩ is recommended for RT. The VCO free running frequency is adjusted with RT and CT to be at the center of input frequency range. It may be seen that PLL is internally broken between the VCO output and phase comparator input.

A short circuit between pins 4 and 5 connects the VCO output to phase comparator, so as to compare f_o with the input signal f_s. A capacitor C is connected between pin 7 and pin 10 to make a LPF with internal resistance of 3.6 KΩ.

2.5.2 Applications

AM Detection

A PLL may be used to demodulate the AM signals.

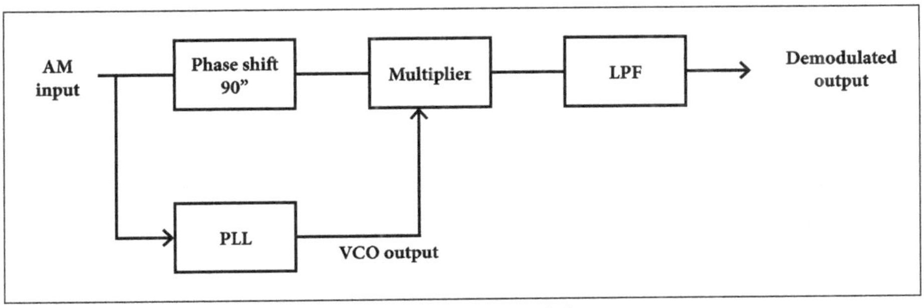

PLL used as AM demodulator.

The PLL is locked to carrier frequency of the incoming AM signal. The output of VCO has the same frequency as the carrier, but unmodulated to fed to multiplier. Since VCO output is always 90° out of phase with the incoming AM signal under the locked condition, the AM input signal is also shifted in phase by 90° before being fed to multiplier. This makes both the signals applied to the multiplier in same phase.

The output of the multiplier contains both the sum and difference signals, the demodulated output is obtained after filtering high frequency components by the LPF. Since the PLL responds only to carrier frequencies which are very close to VCO output, the PLL AM detector exhibits a high degree of selectivity and noise immunity.

FM Detection

If PLL is locked to a FM signal, the VCO tracks the instantaneous frequency of the input signal. The filtered error voltage which controls the VCO and maintain lock with the input signal is the demodulated FM output.

The VCO transfer characteristics determine the linearity of the demodulated output. Since VCO used in IC's. PLL is highly linear, it is possible to realize the highly linear FM demodulators.

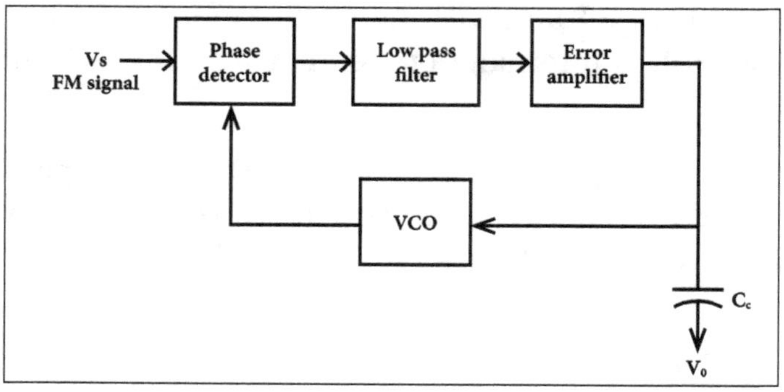

PLL as a FM detector.

FSK

The frequency shift keying (FSK) is a type of frequency modulation, in which the binary data or code is transmitted by means of a carrier frequency that is shifted between two fixed frequency values namely $f_1$representing logic 0 and f_2 representing logic 1.

The frequencies corresponding to logic 0 and logic 1 are called mark and space respectively. The block diagram shown in figure below is a FSK demodulator using PLL.

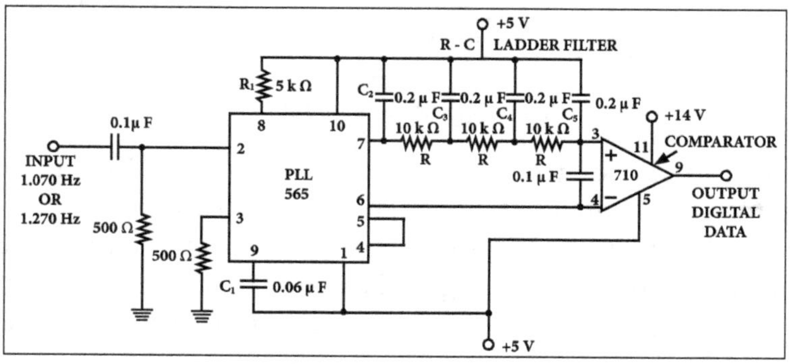

FSK demodulator using PLL.

The PLL is designed to remain in lock, with the FSK signal for both the frequencies f_1 and f_2. Then, the VCO control voltage fed to comparator is given by,

$$V_{f1} = \frac{f_1 - f_o}{K_o}$$

And,

$$V_{f2} = \frac{f_2 - f_o}{K_o}$$

The difference between the two control voltage levels is,

$$\Delta V_f = \frac{f_2 - f_1}{K_o}$$

The two inputs V_{f1} and V_{f2} are applied to comparator. One of the inputs is passed through a second LPF. The filter is designed for a time constant which is longer than the FSK pulse duration to obtain a DC voltage. This DC voltage will have a value mid-way between V_{f1} and V_{f2}.

When the FSK signal V_i is applied to the input, the loop gets locked to the input frequency and tracks it in between the two frequency is f_1 and f_2. The corresponding dc shift at the output is made logic compatible by adjusting the saturation voltages of the voltage comparator.

Frequency Synthesizing

The PLL can be used as basis for frequency synthesizer that can produce a precise series of frequencies that are derived from a stable crystal controlled oscillator. Figure below shows the block diagram of frequency synthesizer.

It is similar to frequency multiplier circuit except that divided by M network is added at the input of phase lock loop. The frequency of the crystal-controlled oscillator is divided by an integer factor M by divider network to produce a frequency, where fOSC is the frequency of the crystal controlled oscillator.

The VCO frequency, f_{VCO} is similarly divided by factor N by divider network to give frequency equal to f_{VCO}/N. When the PLL is locked in on the divided-down oscillator frequency, we will have $fOSC/M - f_{VCO}/N$, so that $f_{VCO} - (N/M)\ f_{OSC}$.

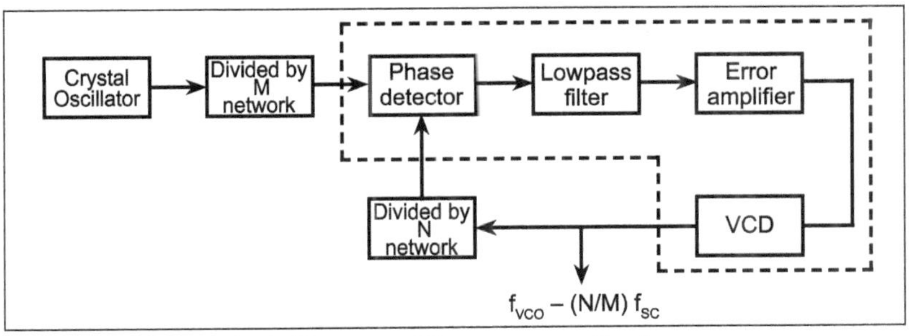

Block diagram of frequency synthesizer.

By adjusting divider counts to desired values, large number of frequencies can be produced, all derived from the crystal controlled oscillator.

3

Data Converters

3.1 Introduction to Data Converters

In general terms a data converter is a device that converts analog signals to digital data or vice versa. Devices that convert analog signals to digital are frequently referred to as ADC. Devices that convert digital signals to analog are called DAC.

Data converters come in many different architectures and are used for a wide variety of different purposes.

Data Converters Uses

DAC and ADC devices each have their own uses both are vital technologies for the digital age. They make many of the luxuries with which people are familiar possible and also play significant roles in scientific, telecommunications and other applications.

Digital to analog converters are mainly used in audio and video applications. They are also used in mechanical applications, but their use in A/V is demonstrative of what these devices offer in terms of the benefits of converting analog signal to digital information.

In music players that use digital media, which constitutes the majority on the market today, digital to analog converters make it possible to actually hear the music. The music is stored in digital format, which is accomplished by using an analog to digital converter, detailed below. The digital information a series of ones and zeroes is take the digital information and convert it back into an analog signal, the voltage that is put through the speakers, producing the sound.

These also play a significant role in VoIP technology. VoIP technology takes analog information the sound of a speaker's voice and samples it and then converts it into digital information. The digital information is transmitted over the line. Before the listener can hear the voice, it is decoded by a DAC, producing the speaker's voice over the line.

DAC technology is also used in video technology, as well, though less heavily than it was in the past. Today, most monitors have digital inputs, which eliminate the necessity to convert the digital data to analog. Should a monitor have analog inputs, however, a DAC is required to convert the information from one form to the other.

Analog to digital converters, intuitively enough, are used for the reverse of the processes listed above, but are also employed in many other applications.

In the case of music, the ADC devices are oftentimes used during the recording process. In order for music to be encoded onto a CD or another digital format, the audio signal has to be sampled and converted to numerical data that can be encoded into the desired format. The ADC provides this capability.

Where ADC devices are concerned, sample rates are vital specs. The sample rate determines how accurately the ADC can represent the analog signal in digital form. The higher the sample rates the better.

ADC devices can work very fast, and that's vital for some signal processing applications. Scientific instruments, such as oscilloscopes and micro controllers commonly employ these devices in their designs. These devices are also used in applications where a signal such as a television broadcast or a radio signal is converted to digital data, such as in the case of TV tuner cards and software defined radio.

ADC technology is also used extensively in scientific instruments of various types. They are used to process signals in some radar systems. They are also used to quantize information recorded as analog signal, such as light levels, temperature, electrical characteristics and other data. This information is read in sensors that produce analog output, which is then converted to digital information that can be read by the operator on a display. These devices commonly include an amplifier for the analog signal.

Difference between a Digital-to-Analog Converter and an Analog-to-Digital Converter

ADAC and an ADC differ in the obvious regard that one converts digital signal to analog and the other analog signal to digital. However, there is a more substantive difference between these two devices in practice.

A digital to analog converter starts with data that is completely and wholly understood. It is represented by a series of 1s and 0s, all its parameters defined. The converter takes that data and converts it into some sort of an analog signal. Because, as long as it is not corrupted digital data can be copied endlessly without changing that data, a DAC has the advantage of always starting out with a known set of values and simply has to convert them into the analog signal.

By contrast, the work of an ADC is much more difficult. Analog signal can be random. The signal is generally continuous with modulations within it that have to be sampled and converted to digital data.

The rate at which the ADC can sample, as well as the resolution of those samples meaning how subtle the differences between one period of the signal and the next can be read by the ADC determine the quality of the digital information the ADC produces.

The ADC is a complex device that has to take into consideration the need for signal conditioning and other methods of improving the information that it works with, thus allowing it to produce a better digital representation of the analog signals being read.

3.2 Basic DAC Techniques, Different Types of DACs, Weighted Resistor DAC, R-2R Ladder DAC and Inverted R-2R DAC

Digital to Analog Convertor (DAC)

The converter which converters the digital form of data in to analog form is called digital to analog converter. In this the digital data in the form of 1's and 0's are used to control the switches which are placed in a analog circuit with reference voltage, based on this switch condition (ON/OFF) and position (MSB or LSB) the output analog amplitude is calculated.

Operation of Digital to Analog Converter

A digital to analog signal converter starts out with a binary number, which is an abstract value that could refer to anything. It takes this information and through the circuitry in the device, converts them to an analog signal that can be put through speakers or another device to achieve the desired result: audio signal, video signal, etc.

To accomplish this, the numbers are converted to impulses. Those impulses are run through a reconstruction filter. This allows the gaps in the data these exist even at the fastest sampling rates to be filled in, producing a more accurate result.

Ideally, the variations in the signal produced by the DAC should be very flowing and very natural. This is particularly important in music and video, where poor sampling rates and audio conversion can result in a choppy, low-fidelity sound or inconsistency in the colour gradients and other aspects of video.

DAC inputs the binary number and outputs the corresponding analog voltage or current signal.

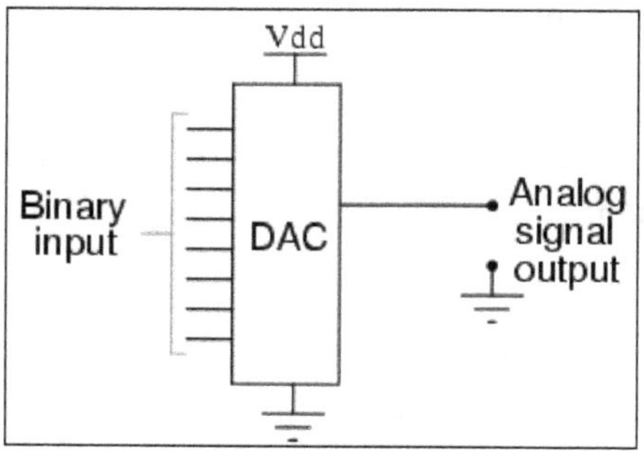

DAC block diagram.

Specifications

Resolution (Output step size):

It is given by the output voltage range divided by the number of possible output levels. An N-bit DAC can output 2^N different levels in 2^N-1 steps. N ranges from 6 (for simple high-speed converters) up to 20 or more (for precision instrumentation).

Accuracy:

It is the maximum difference between the output and the theoretical output. This is approximately the same as the resolution.

Settling time:

It is the time it takes the output to reach a percentage (e.g. 99%) of the final value. Values can range from hundreds of milliseconds for high-precision units to nanoseconds for devices used in high-frequency waveform generators.

Linearity:

It is the maximum difference between the output and a linear interpolation between the minimum and maximum values. This is usually is in the order of $1/2^N$.

Monotonic

The output is monotonic if an increase in the digital input always causes an increase in the analog output. This is usually guaranteed.

There are various types of digital DAC interfaces. Some devices have chip-select and address pins and can be easily interfaced to microcomputers as if they were peripheral chips. Other devices omit the registers and thus require additional logic to interface to a microprocessor.

There are various types of DAC output circuits. Some are designed to generate a current (rather than a voltage) that is proportional to the digital input. Often the DAC is simply a set of N switches and an accurate voltage divider network and an external voltage reference must be supplied.

Finally, there are different ways of manufacturing DAC devices. In addition to the standard monolithic IC's, there are also hybrid devices that mount IC chips together with a small number of discrete devices into an IC-like package.

This allows the manufacturer to adjust ("trim") each device during manufacture to obtain higher precision. There are also DAC modules in which a D/A converter built from discrete components is packaged in a sealed module. Hybrids and modules tend to be used for higher-performance devices.

3.2.1 Different Types of DACs-Weighted Resistor DAC, R-2R Ladder DAC, Inverted R-2R DAC

Types of Digital to Analog Converters:

- Binary Weighted Resistor DAC or R-2nR DAC.
- R-2R Ladder DAC.
- Inverted R-2R DAC.

3.2.2 Weighted Resistor DAC

One of the simplest circuit shown in the below figure uses a summing amplifier with a binary weighted resistor network.

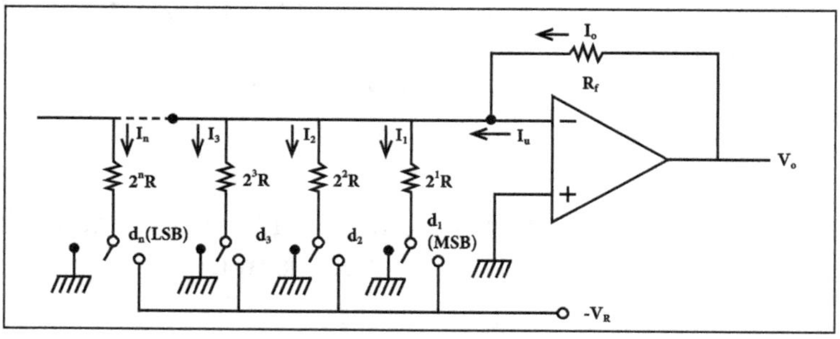

Binary weighted DAC.

It has n-electric switches d_1, d_2.... d_n controlled by binary input word. These switches are single pole double throw (SPDT) type.

If the binary input to a particular switch is 1, it connects the resistance to the reference voltage ($-V_R$). If the input bit is 0, the switch connects the resistor to the ground. In this figure, output current Io for an ideal op-amp can be written as,

$$I_o = I_1 + I_2 + I_3, \ldots. I_n.$$

$$= \frac{V_R}{2R} d_1 + \frac{V_R}{2^2 R} d_2 + \ldots. + \frac{V_R}{2^n R} d_n$$

$$I_o = \frac{V_R}{R} \left[d_1 2^{-1} + d_2 2^{-2} + \ldots\ldots d_n 2^{-n} \right. \qquad \ldots(1)$$

The output voltage,

$$V_o = I_o R_f$$

$$V_o = \frac{V_R}{R} \times R_f \left(d_1 2^{-1} + d_2 2^{-2} + \ldots\ldots + d_n 2^{-n} \right) \qquad \ldots(2)$$

Comparing equations, it can be seen that,

$$R_f = R \text{ and } K = 1 \text{ and } V_{FS} = V_R$$

The analog output voltage is therefore positive staircase as shown in the below figure for a 3-bit weighted resistor.

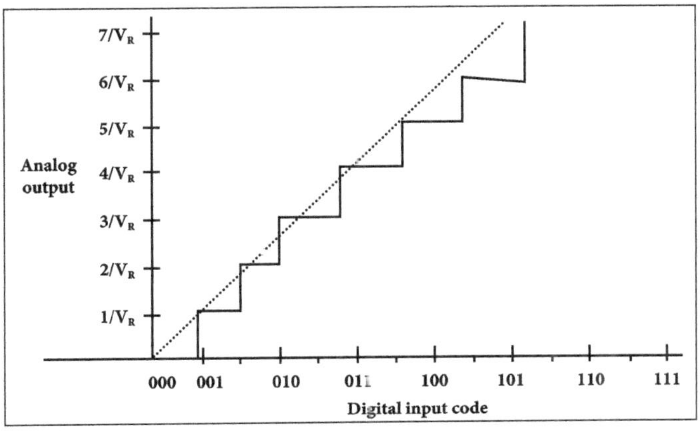

Transfer characteristics of a 3-bit DAC.

Although the op-amp is connected in inverting mode, it can also be connected in non-inverting mode.

The op-amp is simply working as a current to voltage converter.

The polarity of the reference voltage is chosen in accordance with the type of the switch used. For example, for TTL compatible switches, the reference voltage should be +5 V and output will be negative.

The accuracy and stability of a DAC depends upon the accuracy of the resistors and tracking of each other with temperature.

Disadvantages

Wide range of resistor values required. For better resolution, the input binary word length has to be increased. Thus, as the number of bit increases, the range of resistance value increases.

For a 12-bit DAC, the largest resistance required is 5.12 m Ω If the smallest is 2.5 k Ω. The fabrication of such a large resistance in IC is not practical.

The voltage drop across a large resistor due to the bias current would also affect the accuracy. For smaller values of resistors, the loading effect may occur.

The finite resistance of the switches disturbs the binary-weighted relationship among the various currents particularly in the most significant bit positions, where the current setting resistances are smaller.

3.2.3 R-2R Ladder DAC

Wide range of resistors is required in binary weighted resistor type DAC. This can be avoided by using R-2R ladder type DAC where only two values of resistor are required. It is well suited for integrated circuit realization.

The typical value of R ranges from 2.5 k Ω to 10 k Ω.

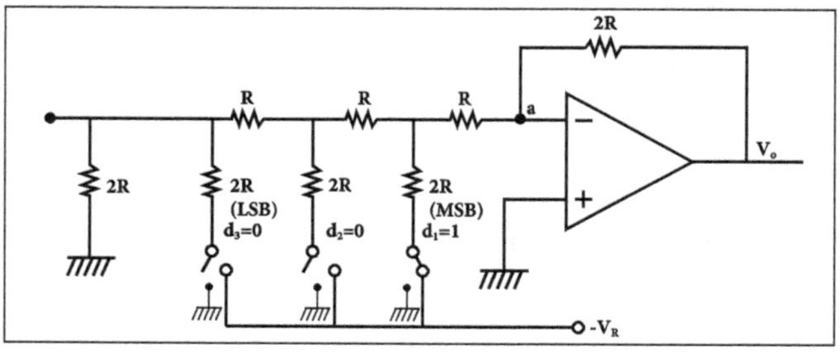

R-2R ladder DAC.

The figure above shows 3-bit DAC where the switch positions d1, d2, d3 corresponds to the binary word 100. The circuit can be simplified to the equivalent form as shown in the below figure:

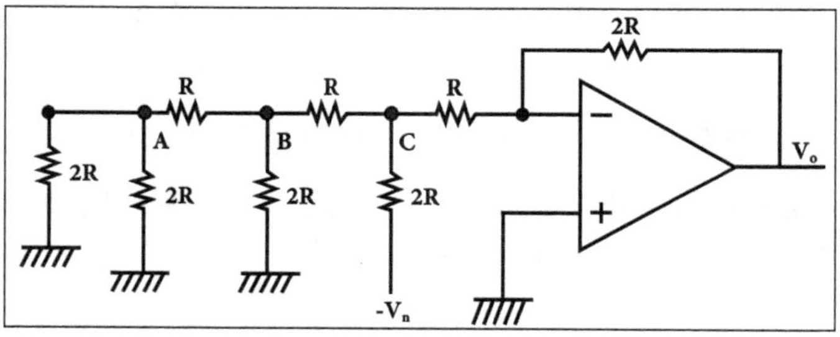

Simplified circuit.

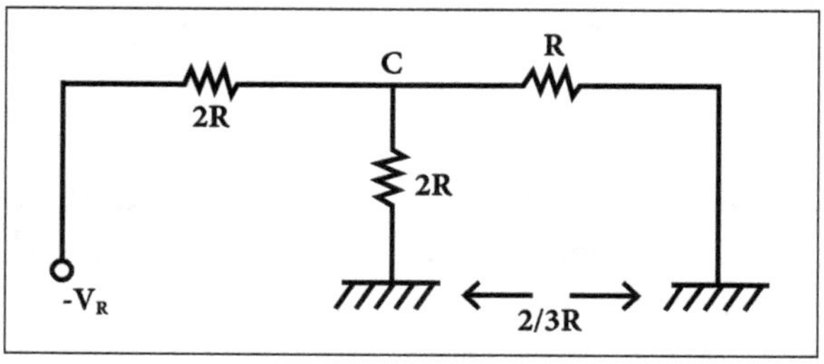

Equivalent circuit.

Then, the voltage at node C can be easily calculated by the set procedure of network analysis,

$$\frac{-V_R\left(\frac{2}{3}R\right)}{2R+\frac{2}{3}R}=\frac{-V_R}{4}$$

The output voltage is given by,

$$V_o = \frac{-2R}{R}\left(\frac{-V_R}{4}\right) = \frac{V_R}{2} = \frac{V_{FS}}{2}$$

The switch position corresponding to the binary word 001 in 3-bit DAC is shown in the below figure:

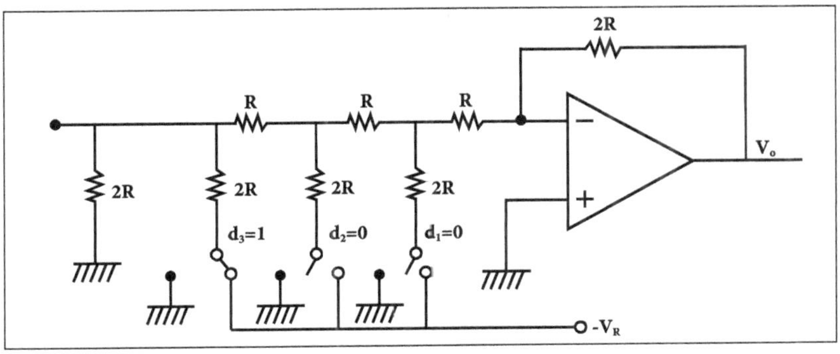

R-2R ladder DAC for switch position 001.

The circuit can be simplified to the equivalent form of the below figure. The voltage at the nodes (A, B, C) formed by resistor branches are easily calculated in a similar fashion and the output voltage becomes,

$$V_o = \left(\frac{-2R}{R}\right)\left(\frac{-V_R}{16}\right) = \frac{V_R}{8} = \frac{V_{FS}}{8}$$

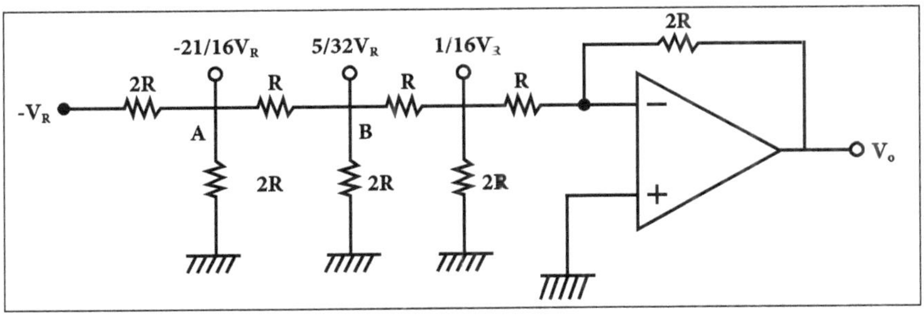

Equivalent circuit.

Similarly, the output voltage for R-2R ladder type DAC corresponding to other 3-bit binary words can be calculated.

Voltage Mode DAC

This is the alternative mode of DAC and hence it is called so because, its 2R resistance in the shunt path is switched between two voltages named as VL and VH. The output of this DAC is obtained from the leftmost ladder node.

As the input is sequenced through all the possible binary state starting from all 0s

(0.....0) to all 1s (1.....1). The voltage of this node changes in steps of 2^{-n} (V_H - V_L) from the minimum voltage of $V_o = V_L$ to the maximum of $V_o = V_H - 2^{-n} (V_H - V_L)$.

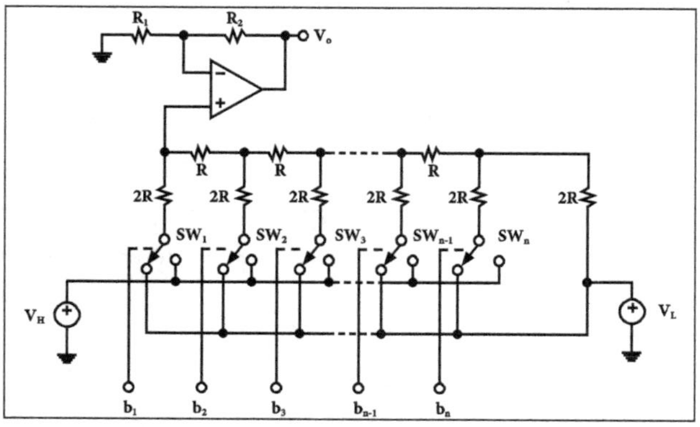

Voltage mode DAC.

The above figure also shows a non-inverting amplifier from which the final output is taken. Due to this buffering with a non-inverting amplifier, scaling factor is defined by $K=1+(R_2/R_1)$ results.

Advantages

- The major advantage of this technique is that it can allow us to interpolate between any two voltages, which need not be a zero.
- More accurate selection and design of resistors R and 2R are possible and simple construction.
- The binary word length can be easily increased by adding the required number or R-2R sections.

Current Mode DAC

Current mode DAC's operates based on the ladder currents. The ladder is formed by resistance R in the series path and resistance 2R in the shunt path. Thus, the current is divided into i_1, i_2, i_3,...... in in each arm. The currents are either diverted to the ground bus (io) or to the virtual ground bus $\left(\bar{i}_o\right)$.

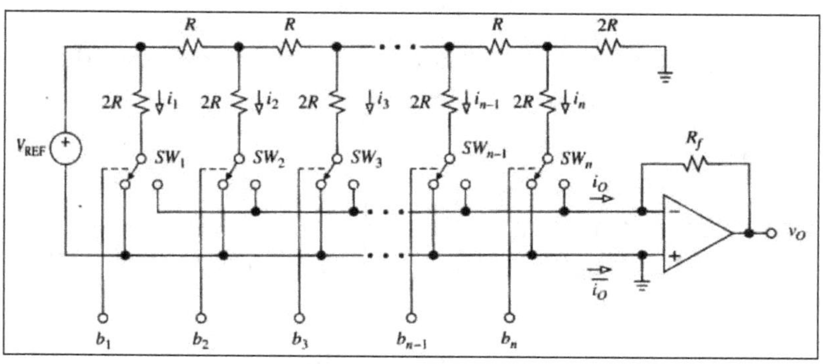

Current Mode DAC.

The currents are given as,

$$i_1 = V_{REF}/2R = (V_{REF}/R)_2^{-1}$$

$$i_2 = (V_{REF})/2)/2R = (V_{REF}/R)_2^{-2}$$

$$\vdots$$

$$i_n = (V_{REF}/R)_2^{-n}$$

And the relationship between the currents are given as,

$$i_2 = i_1/2$$

$$i_3 = i_1/4$$

$$i_4 = i_1/8$$

$$i_n = i_1/2^{n-1}$$

Using the bits to identify the status of the switches and letting $V_o = -R_f i_o$ gives,

$$V_o = -(R_f/R)\, V_{REF}\left(b_1 2^{-1} + b_2 2^{-2} + \ldots\ldots + b_n 2^{-n}\right)$$

The two currents i_o and $\overline{i_o}$ are complementary to each other and the potential of i_o bus must be sufficiently close to that of the $\overline{i_o}$ bus. Otherwise, linearity errors will occur. The final op-amp is used as current to voltage converter.

Advantages

In the Current mode or inverted ladder type DAC's, the stray capacitance do not affect the speed of the response of the circuit due to constant ladder node voltages. Thus, it improves the speed performance.

The major advantage of the current mode D/A converter is that the voltage change across each switch is minimal. Hence, the charge injection is virtually eliminated and the switch driver design is made simpler.

3.2.4 Inverted R-2R DAC

The following R-2R ladder network figure, has the following points,

The binary bit positions of the digital inputs are interchanged.

Each of the binary inputs Vin(0), Vin(1), Vin(2), and Vin is connected through the resistors to one end of the corresponding switches. The other end of these switches is connected to the negative input terminal of the op amp when the binary bit is in one (1) state.

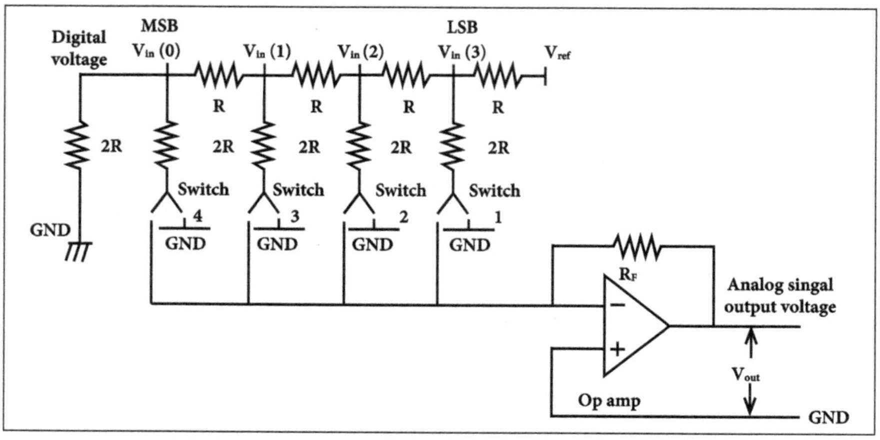

Digital to analog converter using inverter R-2R ladder.

The negative input terminal of the op amp is also at virtual ground (GND in the figure) for practical purposes.

The other end of the resistors is connected to the ground terminal when the binary bit is in zero (0) state.

The switching nodes in the two positions are at ground potentials. Hence, the parasitic capacitances associated with switching are minimized, and consequently, this DAC works faster than the other types of DAC3zs.

The four input voltages can be considered as follows,

$$V_{in}(0)=D_1$$

$$V_{in}(1)=D_2$$

$$V_{in}(2)=D_3$$

$$V_{in}(3)=D_4$$

Then output voltage $V_{out} = V_{ref} \times \left[\frac{D_1}{16} + \frac{D_2}{8} + \frac{D_3}{4} + \frac{D_4}{2} + \ldots\right]$ Volts

IC 1408 DAC

The 1408 is an 8 bit R/2R ladder type D/A converter compatible with rn and CMOS logic. It is designed to use where the output current is linear product of an eight bit digital word.

The below figure (a), shows the pin diagram and block diagram for IC 1408 DAC.

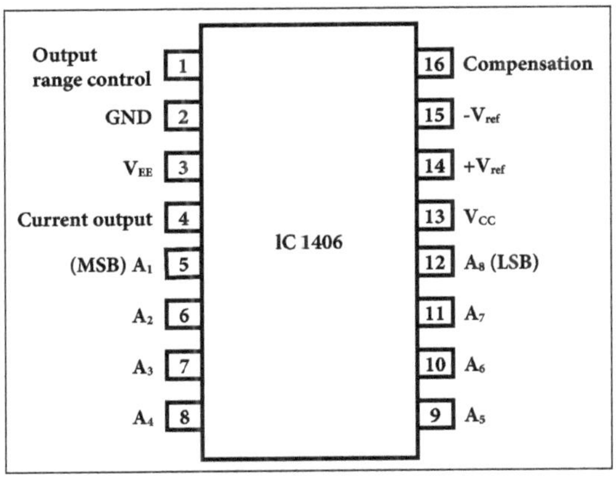

(a) Pin diagram.

The IC 1408 consists of a reference current amplifier, an R/2R ladder and eight high speed current switches. It has eight input data lines A1 (MSB) through A8 (LSB) which control the positions of current switches.

It requires 2 mA reference current for full scale input and two power supplies $V_{cc} = +5$ V and $V_{EE} = -15$ V (V_{EE} can range from - 5 V to - 15 V).

The voltage V_{ref} and resistor R_{14} determines the total reference current source and R15 is generally equal to R14 to match the input impedance of the reference current amplifier.

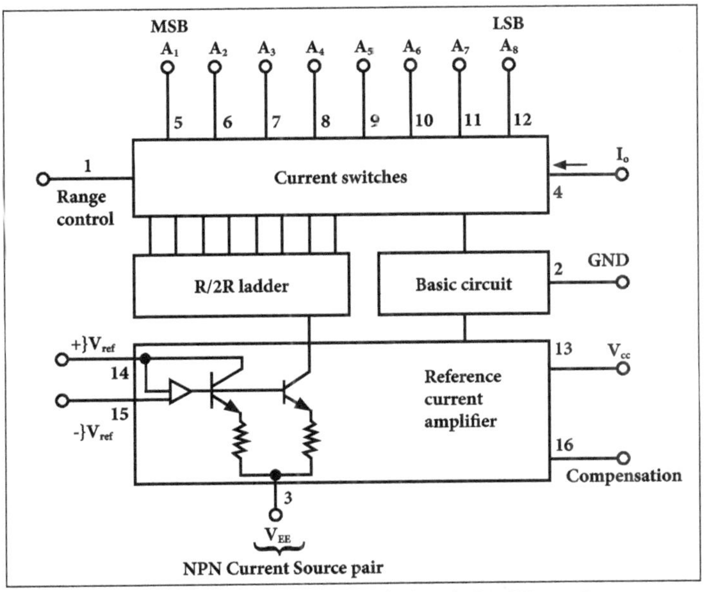

Block diagram for a typical circuit for IC 1408.

The output current IO can be given by,

$$I_O = \frac{V_{ref}}{R_{14}}\left(\frac{A_1}{2} + \frac{A_2}{4} + \frac{A_3}{8} + \frac{A_4}{16} + \frac{A_5}{32} + \frac{A_6}{64} + \frac{A_7}{128} + \frac{A_8}{256}\right)$$

Note: input A1 through A_8 can be either 0 or 1. Therefore, typical circuit full scale current can be given by,

$$I_O = \frac{5}{2.5K}\left(\frac{1}{2}+\frac{1}{4}+\frac{1}{8}+\frac{1}{16}+\frac{1}{32}+\frac{1}{64}+\frac{1}{128}+\frac{1}{256}\right)$$

$$= \frac{2\,mA \times 255}{256}$$

$$= 1.992\,mA$$

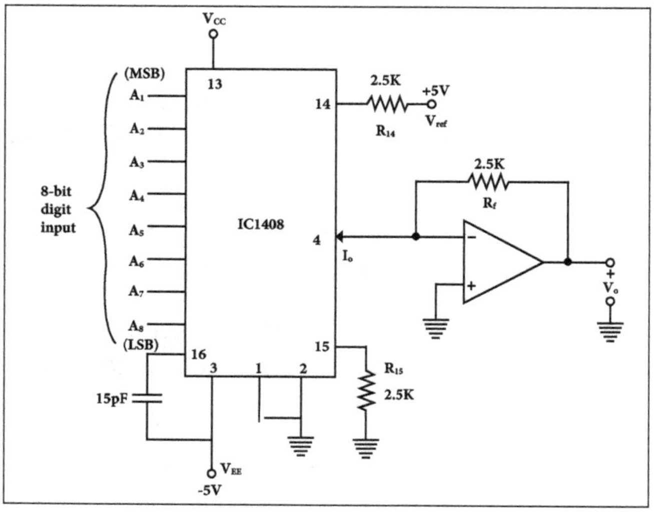

Typical circuit for IC 1408.

It shows that the full scale output current is always 1 LSB less than the reference current source of 2 mA. This output current is converted into voltage by I to V converter. The output voltage for full scale input can be given as,

$$V_o = 1.992 \times 2.5K$$

$$= 4.98\ V$$

Important Electrical Characteristics for IC 1408:

- Reference current: 2 mA.
- Supply voltage: + 5 Vcc and – 15 V WE.
- Setting time: 300 ns.
- Full scale output current: 1.992 mA.
- Accuracy: 0.19 %.

3.3 Different Types of ADCs

Operation of Analog to Digital Converter

An analog to digital converter is a complex device. The process of converting the signal involves taking small amounts of data samples out of an analog signal. These samples are converted to discrete values the 1s and 0s that are used to store a digital representation of the analog signal sampled or to output that data to a display.

For analog signals, resolution is a significant determiner of how well the device represents the signal in digital form.

The inherent challenge for an ADC is the fact that analog signal has essentially, the potential for an infinite amount of variations in voltage within the signal. The ADC takes the signal recorded in a sample and compares it to a reference voltage. This is used to make the conversion from analog to digital.

An important aspect of ADC devices is quantization error. This is an inevitable problem with ADC technology. Because only portions of the analog signal are actually recorded and converted, there is always some lost signal during the analog to digital conversion process. The highest quality ADC technology loses very little, but there are losses, nonetheless.

Once the ADC has sampled the analog signal, it takes that sample and records it as digital data, adding to it as more signals is sampled and converted to binary numbers.

A/D Converters

An ADC inputs an analog signal such as current or voltage and it outputs the digital signal in the form of a binary number.

Specifications

Accuracy:

Accuracy is the total error with which the A/D converter can convert a known voltage, including the effects of quantization error, gain error, offset error and nonlinearities. Technically, accuracy should be traceable to known standards (for example, NIST) and is generally a catch-all term for all static errors.

Resolution:

Resolution is the number of bits, N, out of the A/D converter. Resolution determines the size of the least significant bit and thus determines the dynamic range, the code widths and the quantization error.

Dynamic Range

Dynamic range is the ratio of the smallest possible output (the least significant bit or quantum voltage) to the largest possible output (full-scale voltage), mathematically $20 \log_{10} 2^N \approx 6N$.

Offset Error

Offset error is the deviation in the A/D converter's behavior at zero. The first transition voltage should be 1/2 LSB above analog ground. Offset error is the deviation of the actual transition voltage from the ideal 1/2 LSB. Offset error is easily trimmed by calibration.

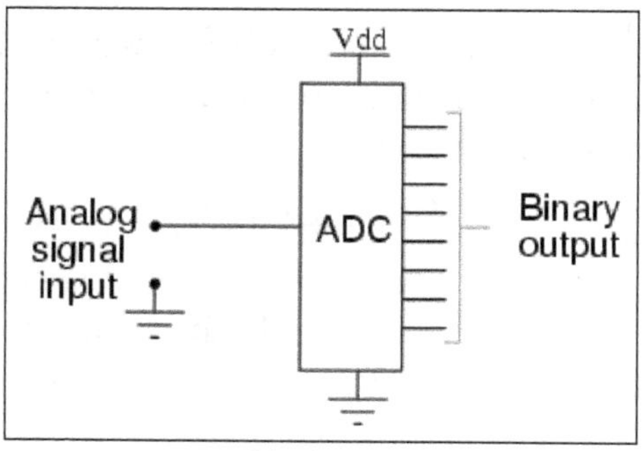

ADC block diagram.

Gain Error

Gain error is the deviation in the slope of the line through the A/D converter's end points at zero and full scale from the ideal slope of 2N/VFS codes-per-volt. Like offset error, gain error is easily corrected by calibration.

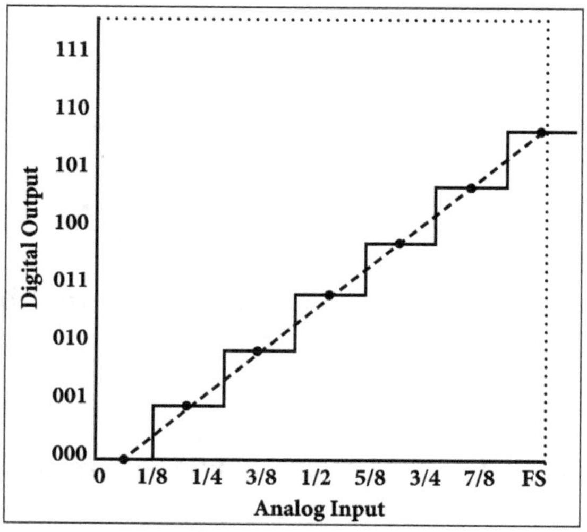

Offset and gain error.

Differential Non-Linearity

Differential non-linearity (DNL) is the deviation of the code transition widths from the ideal width of 1 LSB. All code widths in the ideal A/D converter are 1 LSB wide, so the DNL would be zero everywhere. Some data sheets list only the "maximum DNL". The wide range of code widths are shown in the below figure:

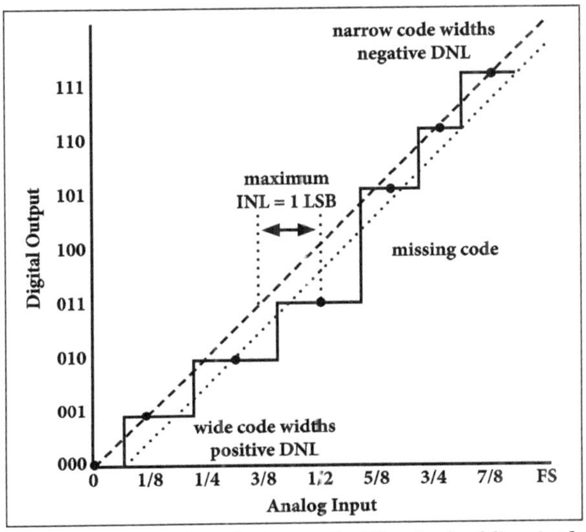

Nonlinearity errors and a missing code. The dashed line is the ideal characteristic and the dotted line is the best fit.

Integral Non-Linearity

Integral non-linearity (INL) is the distance of the code centers in the A/D converter characteristic from the ideal line. If all code centers land on the ideal line, the INL is zero everywhere. Some data sheets list only the maximum INL. The deviations of the code centers from the ideal line is shown in the above figure.

Note that there are two possible ways to express maximum INL, depending on the definition of the ideal line. Data sheet INL numbers can be decreased by quoting maximum INL from a "best fit" line instead of the ideal line.

In the above figure, the ideal line (shown dashed) exhibits an INL of 1 LSB, while the best-fit line (shown dotted) exhibits an INL of half that size. While this prevarication underestimates the effect of INL on total accuracy, it is probably a better reflection of the A/D converter's linearity in AC-coupled applications.

Missing Codes

Missing codes are output digital codes that are not produced for any input voltage, usually due to large DNL. In some converters, missing codes can be caused by non-monotonicity of the internal D/A. The large DNL in the above figure causes code 100 to be crowded out.

Different Types of ADCs

Types of A/D converters are:

- Flash (The parallel comparator) type A/D converter.
- Successive approximation type.
- Single slope type.
- Dual slope type.

3.3.1 Parallel Comparator Type ADC

This is the simplest possible A/D converter. Fastest and most expensive technique. The below figure shows a 3-bit A/D converter:

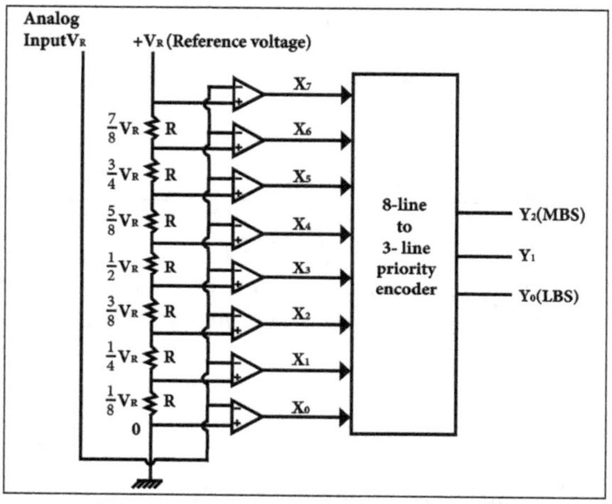

Basic circuit of a flash type A/D converter.

The circuit consists of a resistive divider network, 8 op-amp comparators and a 8 line to 3 line encoders (3-bit priority encoder).

The comparator and its truth table are shown below:

Voltage input	Logic output X
$V_a > Vd$	X = 1
$V_a > Vd$	X = 0
$V_a = Vd$	Previous Value

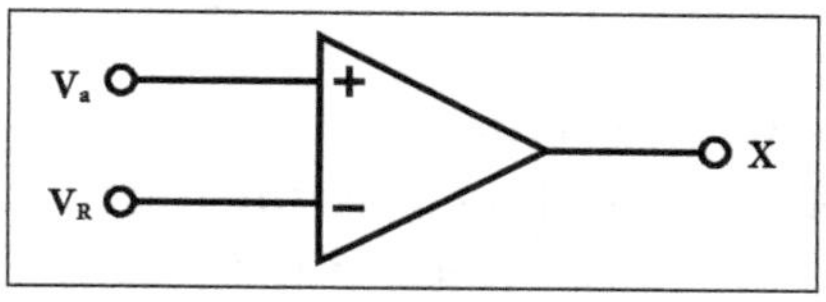

Comparator and its truth table.

Input Voltage V_a	X_7	X_6	X_5	X_4	X_3	X_2	X_1	X_0	Y_2	Y_1	Y_0
0 to $\frac{V_R}{8}$	0	0	0	0	0	0	0	1	0	0	0
$\frac{V_R}{8}$ to $\frac{V_R}{4}$	0	0	0	0	0	0	1	1	0	0	1
$\frac{V_R}{4}$ to $\frac{3V_R}{8}$	0	0	0	0	0	1	1	1	0	1	0
$\frac{3V_R}{8}$ to $\frac{V_R}{2}$	0	0	0	0	1	0	0	0	0	1	1
$\frac{V_R}{2}$ to $\frac{5V_R}{8}$	0	0	0	1	0	0	0	0	1	0	0
$\frac{5V_R}{8}$ to $\frac{3V_R}{4}$	0	0	1	1	1	1	1	1	1	0	1
$\frac{3V_R}{4}$ to $\frac{7V_R}{8}$	0	1	1	1	1	1	1	1	1	1	0
$\frac{7V_R}{8}$ to V_R	1	1	1	1	1	1	1	1	1	1	1

At each node of the resistive divider, voltage available at the nodes are equally divided between the reference voltage VR and the ground. The purpose of the circuit is to compare the analog input voltage Va with each of the node voltage. The truth table for the flash type A/D converter is shown in the table above.

The circuit has the advantages of high speed as the conversion takes place simultaneously rather than sequentially. Typical conversion time is 100 ns or less. Conversion time is limited only by the speed of the comparator and of the priority encoder. By using an advanced micro devices AMD686A comparator and a T1147 priority encoder, conversion delays of the order of 20 ns can be obtained.

Disadvantages

More number of comparators required.

A 2-bit ADC requires 3 comparators, (2^n-1) comparator. 3-bit ADC needs 7, whereas 4-bit require 15 comparator. In general, no. of comparators required are 2^n-1, where n is the derived number of bits.

Larger the value of n, more complex is the priority encoder.

3.3.2 Counter Type ADC

The following are the components of counter type ADCs:

- Analog input signal (for conversion to digital signal).
- Op amp comparator.
- AND gate.
- Clock generator.
- DAC.
- Reference voltage.
- Binary counter.
- Digital output.

Digital display: The schematic block diagram of a counter type ADC is shown in the below figure. Its input is an analog signal and output is a digital signal, which is a collection of 0s and 1s.

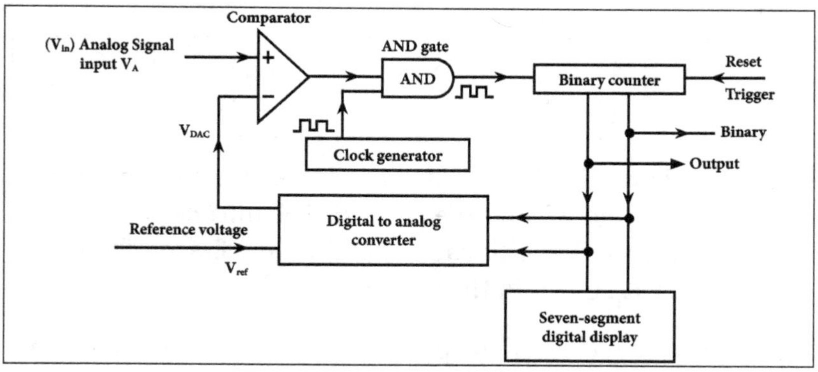

Components of counter type ADCs.

Principle of Operation

- Reset the binary counter to zero (before starting the analog to digital conversion).
- At the beginning of the ADC process, voltage $V_{ac(VD)} = 0$.
- There are two input signals to the op amp comparator.
- The input signal of the ADC is connected to the input terminal of the comparator.
- The output voltage VDAC of the DAC is compared with the analog input signal of the ADC.

- The input signals to the AND gate are the output voltage from the comparator and the clock generator output (clock pulses).
- If the analog input signal Vin is greater than 0, its comparator output will be logical 1. The logical I output enables the AND gate for transmission of clock pulses to the binary counter. The counter then receives the clock pulses.
- The clock pulses from the AND gate are connected to the binary counter.
- The binary counter output is the digital output. It will become 0001.
- The maximum number of count levels N = 2n where n is the number of bits in the counter.
- The digital output from the binary counter is converted to analog voltage VDAC by the DAC and produces an output voltage VDAC = I.

If the analog signal is greater than one at this instant of time, the comparator output goes to logical 1. It enables the AND gate. And the counter advances and reads 0010. The output of the DAC is continuously compared with the analog input (to ADC) for conversion into digital output display.

The comparison of the analog signals V. and VDAC (output of DAC) is made by the comparator. If its output is greater than one. Counting progresses. When the analog input signal equals or is less than the DA output, the counter stops. The digital output is used for the external process.

The binary counter output is the digital output, which is the converted signal of the external analog input signal. Thus, analog to digital signal conversion takes place. The digital output is displayed in the seven-segment display.

Thus, the following are the basic operations involved in the working of a counter type ADC:

- Reset and start counter.
- The DAC converts the digital output of the counter to the analog voltage VDAC.
- The analog input signal voltages Vin and VDAC are compared by the comparator.
- If the comparator produces logical I output voltage, it operates the counter.
- The output of the counter is the digital output of the analog voltage used for the conversion.
- The cycle of events during the ADC conversion process repeat.
- When the analog input signal equals or is less than the DAC output, the counter stops.

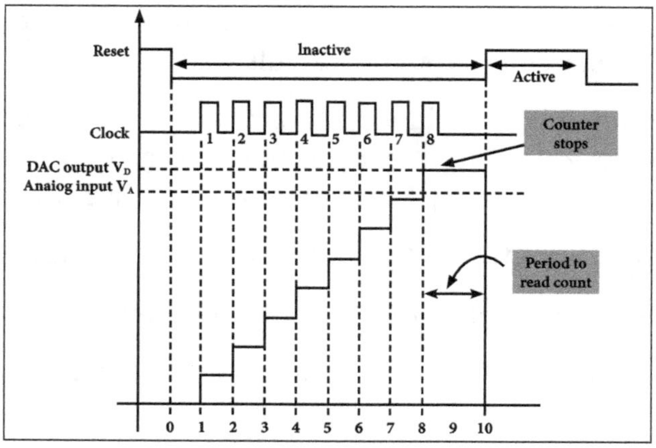

Waveform for counter type ADC.

Advantages

- The operation is simple and straightforward.
- Computer data acquisition system.

Disadvantages

- It is necessary to give enough time for DAC conversion and comparator to respond. Therefore, there is a limitation on the clock frequency. As dock frequency is low, the speed of conversion is less.
- Conversion time is not constant It teases with increase in input voltage. In other words, we can say that conversion time is high at high input voltage.

3.3.3 Successive Approximation ADC

The successive approximation technique uses a very efficient code search strategy to complete n-bit conversion in just n clock periods.

An eight bit converter would require 8 clock pulses to obtain a digital output. The figure below shows an 8-bit converter. The circuit uses a successive approximation register (SAR) to find the EOC required value of each bit by trial and error.

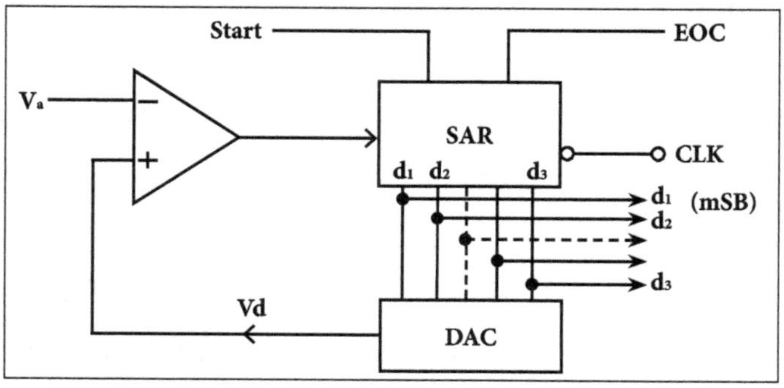

Functional diagram of the successive approximation ADC.

Operation

With the arrival of the START command, the SAR sets the MSB d1 = 1 with all other bits to zero so that the trial code is 10000000.

The output V_d of DAC is now compared with the analog output V_a. If V_a is greater than the DAC output V_d, then 1,00,00000 is less than the correct digital representation.

The MSB is left at 1 and the next lower significant bit is made 1 and further tested. However, if V_a is less than DAC output, then 10000000 is greater than the correct digital representation. So reset MSB to 0 and go on to the next significant bit. This procedure is repeated for all subsequent bits, one at a time until all bit position have been tested.

When the DAC output crosses V_a, the comparator changes state and this can be taken as the End of Conversion (EOC) command. Table below shows a typical conversion sequence.

Successive approximation conversion sequence for a typical input:

Correct digital representation	Successive approximation register output Vd at different stages in the conversion	Comparator output
11010100	10000000	1 [initial output]
	11000000	1
	11100000	0
	11010000	1
	11011000	0
	11010100	1
	11010110	0
	11010101	0

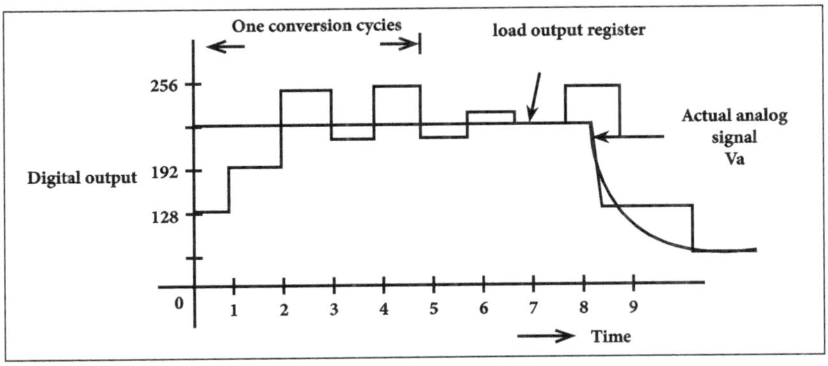

D/A output characteristics.

The figure above shows that the D/A output voltage is seen to become successively closer to the actual analog input voltage.

It can be seen that the D/A output voltage becomes successively closer to the actual input voltage. It requires 8 pulses to establish the accurate output regardless of the value of the additional analog input. However, one circuit pulse is used to load the output register and reinitialize the circuit.

3.3.4 Dual Slope ADC

Dual slope conversion is an indirect method for A/D conversion where an analog voltage and reference voltage are converted into time periods by an integrator and then measured by counter.

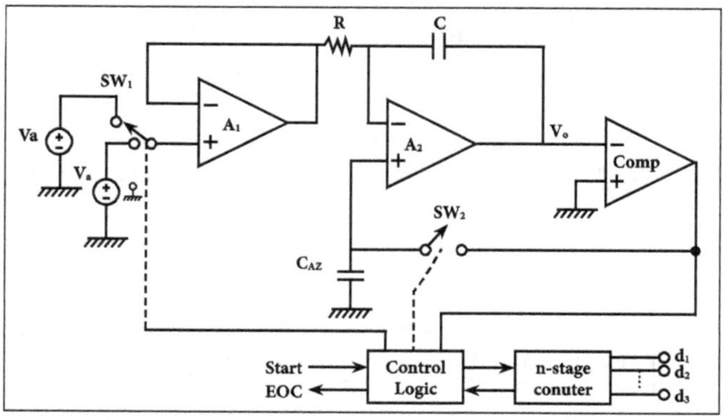

Functional diagram of the dual slope ADC.

The above figure shows the functional diagram of the dual slope or dual ramp converter. The analog part of the circuit consists of a high input impedance buffer A1, precision integrator A2 and a voltage comparator.

The converter first integrates the analog input signal for a fixed duration of 2n clock periods as shown in the below figure. Then, it integrates an internal reference voltage VR of opposite polarity until the integrator output is zero.

The number N of clock cycles required to reduce the integrator to zero is proportional to the value of V_a averaged over the integration period. Hence N represents the desired output code.

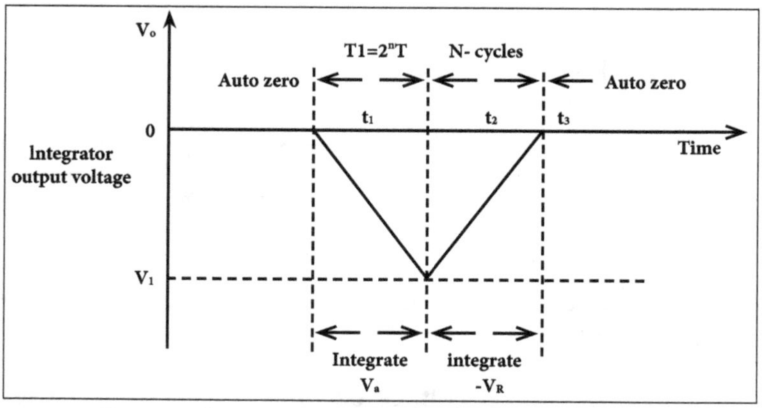

Integrated output waveform for the dual slope ADC.

Operation

Before the START command arrives, the switch SW_1 is connected to ground and SW_2 is closed. Any offset voltage present in the A_1, A_2 and comparator loop after integration appears across the capacitor till the threshold of the comparator is achieved.

The capacitor C_{AZ} thus provides automatic compensation for the input offset voltages of all the three amplifier. Later, when SW_2 opens C_{AZ} acts as a memory to hold the voltage required to keep the offset null. At the arrival of the START command at $t = t_1$, the control logic opens SW_2 and connects SW_1 to Va and enables the counter starting from zero. This circuit uses an n-stage ripple counter and therefore the counter resets to zero after counting 2^n pulses.

The analog voltage V_a is integrated for a fixed number 2^n counts of clock pulses after which the counter resets to zero. If the clock period is T, the integration takes place for a time $T_1 = 2^n \times T$ and the output is a ramp going downwards as shown in the above figure.

The counter resets itself to zero at the end of the internal T1 and the switch SW_1 is connected to the reference voltage (-VR). The output V_o will now have a positive slope. As long as V_o is negative, the output of the comparator is positive and the control logic allows the clock pulse to be counted.

However, when V_o becomes just zero at time $t = t_3$, the control logic issues an EOC command and no further clock pulses enter the counter. It can be shown that the reading of the counter at t_3 is proportional to the analog output voltage V_a.

In the figure,

$$T_1 = t_2 - t_1 = \frac{2^n \text{ counts}}{\text{Clock rate}}$$

$$t_3 - t_2 = \frac{\text{digital count N}}{\text{Clock rate}}$$

For an integrator,

$$\Delta V_o = \left(-\frac{1}{RC}\right) V(\Delta t)$$

The voltage V_o will be equal to V_1 at the instant t_2 and can be written as,

$$V_1 = \left(-\frac{1}{RC}\right) V_a (t_2 - t_1)$$

The voltage V_1 is also given by,

$$V_1 = \left(-\frac{1}{R_C}\right) - V_R\left(t_2 - t_3\right)$$

$$V_a\left(t_2 - t_1\right) = V_R\left(t_3 - t_2\right)$$

Putting the values of $\left(t_2 - t_1\right) = 2^n$ and $\left(t_3 - t_2\right) = N$,

We get,

$$V_a\left(2^n\right) = V_R\left(N\right)$$

$$V_a = \left(V_R\right)\left(N/2^n\right)$$

Since V_R and n are constant, the analog voltage V_a is proportional to the count reading N and is independent of R, C and T.

The dual slope ADC integrates the input signal for a fixed time. Hence, it provides excellent noise rejection of AC signals whose periods are integral multiples of the integration time T_1.

The main disadvantage of the dual slope ADC is the long conversion time. Dual slope converters are particularly suitable for accurate measurement of slowly varying signals such as thermocouples and weighing scales. Dual slope ADC's also form the basis of digital panel meter and multi-meter.

3.4 DAC and ADC Specifications

For better ESDM processes, the inner circuit details of the different types of ADC ICs and DAC ICs are to be understood clearly. Before the selection of an IC, its specifications are more useful for the system design.

1. Temperature range and electrical and mechanical specifications for the following types of applications:

- Commercial.
- Industrial.
- Military.

2. Type of IC package and number of pins:

- Dual inline package.

- Metal casing.
- Plastic.

3. Architecture of the internal circuit:

- Dual slope integrator type ADC.
- R-2R ladder type network.
- Parallel (flash).

4. Interface type at the output port of the IC:

- Serial port interface.
- Parallel port interface.

5. Input terminal types:

- Single-ended pair of input terminals.
- Differential input terminals.

Multiple inputs:

6. Sampling rates of input signals -555 KHz (typical value).

7. Operating voltages and currents:

- Operating voltages: 3 V or 5 V.
- Operating current. 1.25 mA (typical value).

8. Supply voltage variations.

9. Power dissipation.

10. Temperature variations, as they affect convener performance features.

11. Interfacing capability with other devices.

12. Resolution

13. Conversion time, which is the total time taken for the analog to digital signal con-version and propagation delays in the total circuit of the ADC.

14. Linearity

15. Number of binary digital display for the ADC-4-bit display (typical value).

16. Zero offset.

17. Accuracy, which is the difference between the expected output and the actual output (e.g., in DAC, accuracy will be defined in terms of the difference between the estimated analog output voltage and the measured output voltage)

18. Quantization error, which is associated only with AD (' circuits.

19. Settling time in high-speed data converters, which is the time taken for the output voltage to settle down at the final response within a specified error band for step voltage input and is caused by propagation delay.

20. Stability.

Specifications AD 574 (12 bit ADC)

The AD574A is a complete 12-bit A/D converter which requires no external components to provide the complete successive approximation analog-to-digital conversion function. The specifications of A17574AJ are shown in the below table:

Parameter	value
Resolution	12-bit
Linearity error	± 1LSB
Temperature range	0- 70° C
Analog input range	-5 to 5v
Power supply:	
V_{logic}	+4.5 to +5.6
V_{cc}	+11.4 to +16.5
V_{EE}	-11.4 to -16.5
Power dissipation	390 mW
	725 mW
Internal reference voltage	9.98 to 10.02
Maximum conversion time	35μs
Operating current:	
I_{LOGIC}	30 to40mA
I_{cc}	2-5 mA
I_{EE}	18 to 30 mA

4

Digital Integrated Circuits

4.1 Classification of Integrated Circuits

Integrated circuits offer a wide range of applications and could be broadly classified as:

- Digital ICs.
- Linear ICs.

Based upon the above requirements, two distinctly different IC technologies namely, Monolithic technology and Hybrid technology have been developed.

In monolithic integrated circuits, all circuit components, both active and passive elements and their interconnections are manufactured into or on top of a single chip of silicon. The monolithic circuit is ideal for applications where identical circuits are required in very large quantities and hence provides lowest per-unit cost and highest order of reliability.

In hybrid circuits, separate component parts are attached to a ceramic substrate and interconnected by means of either metallization pattern or wire bonds. This technology is more adaptable to small quantity custom circuits. Based upon the active devices used, ICs can be classified as bipolar (using BJT) and unipolar (using FET). Bipolar and unipolar ICs may further be classified depending upon the isolation technique or type of FET used as in the below figure:

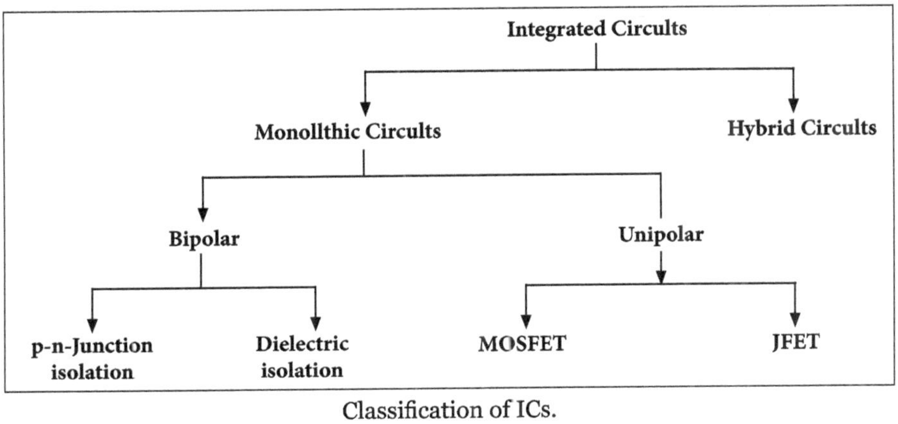

Classification of ICs.

Comparison of Various Logic Families Combinational Logic ICs:

Logic Family Or Series	Logic Technology	Supply Voltage (Volts)	Basic Details
RTL	Resistor Transistor Logic	~3	The RTL logic technology was one of the very first introduced. It first was available in 1962 and as the name implies it used just resistor and transistors.
DTL	Diode transistor logic	5	The DTL logic family or series used diodes on the inputs to create functions such as AND and OR and then had transistors on the output. It was soon superseded by the later TTL logic technology.
MECL I	Emitter coupled logic	–5.2	ECL was the first type of logic IC family to be developed and it appeared around 1962.
ECL III	Emitter coupled logic	–5.2 (-5.19 - - 5.21)	This was an improved form of ECL emitter coupled logic. It was very fast having toggle speeds of up to 500 MHz, but it was very current hungry and chips would run very warm. However its speed meant that It remained in use. Often being used in RF applications such as frequency synthesizers where very fast counters were needed. Running from a negative supply it was difficult to interface to many other elements of the design although interface chips were available.
ECL 10K	Emitter coupled logic	-5.2 (-5.19 - - 5.21)	An updated version of ECL, although it had a top toggle speed of around 125 MHz.
ECL 100K	Emitter coupled logic	-5.2 (-5.19 - - 5.21)	An even later version of ECL but with a maximum clock speed of 350 MHz.
7400	TTL	5 (4.75 – 5.25)	Introduced around 1964, this series of logic used transistors in all areas of the chip. It became an industry standard. This technology and its numbering scheme were adopted for many future generations of logic chip technology. The numbers for the devices were retained such that the function contained within a 74**XX chip would remain the same across the different logic technologies.

74L00	TTL (LOW POWER)	5 (4.75 – 5.25)	Introduced In 1964, the low power 7400 series of logic devices used lower power than the standard TTL family, but at the expense of speed and output capability.
74H00	TTL (High speed)	5 (4.75 – 5.25)	Again introduced In 1964, this high speed TTL offered a speed of almost twice that of the standard 7400 logic family, but at the expense of increased current consumption.
74S00	TTL (Schottky)	5 (4.75 – 5.25)	Using Schottky diode technology, this form of TTL was able to offer operating speeds of around four times that of the standard 7400 logic family.
74LS00	TTL (LOW POWER Schottky)	5 (4.75 – 5.25)	The low power Schottky logic family offered lower propagation delays, higher toggle speeds and lower power levels per gate than the original 7400 logic family. As such it became very widely used for many years.
74ALS	TTL (Advanced low power Schottky)	5 (4.5 – 5.5)	Advanced low power Schottky giving increased speed and lower power when compared to low power Schottky.
74F00	TTL (Fast)	5 (4.75 – 5.25)	Fast TTL, offering speeds that are four times that of standard 7400 series TTL. While the speed is essentially the same as Schottky TTL, it provides the speed at around a quarter of the power.

4.2 Specifications and Applications of TTL-74XX

The 7400 series of TTL chips became one of the de-facto standards for digital or logic integrated circuits.

Although introduced in the mid-1960s, the basic 7400 TTL chip family went on to form the basis of many of the families which are used to this day. Many elements of the 7400 TTL family of chips remain the same, from the numbering scheme to the basic concepts and standards.

Specifications of TTL - 74XX

Some of the main or highlight features and specifications for the 7400 series logic family are detailed below:

PARAMETER	SPECIFICATION
Supply voltage	Nominal 5V (4.75 – 5.25)
Propagation delay per gate	Typically 10 ns
Max toggle speed	25 MHz
Power consumption per gate	10 mW

Applications of TTL-74XX:

- Used in controller application for providing 0 to 5Vs.
- Used as switching device in driving lamps and relays.
- Used in processors of mini computers like DEC VAX.
- Used in printers and video display terminals.

4.3 Code Converters, Decoders, DE-Multiplexers, LED and LCD Decoders with Drivers

Code Converter

Numbers are usually coded in one form or another so as to represent or use it as required. For instance, a number nine is coded in decimal using symbol (9) d. Same is coded in natural-binary as (1001)b.

While digital computers deal with binary numbers there are situations wherein natural-binary representation of numbers are in-convenient or in-efficient and some other (binary) code must be used to process the numbers.

Binary to Gray Code Converter

Gray code is non-weighted code. It has a very special feature that only one bit will change.

A binary number can be converted to gray by following steps:

- Record the MSB as it is.
- Add this bit to the next position, recording the sum and neglecting any carry.

- Record successive sums until completed.

Application of Gray Code:

- Gray code is used in applications where the normal sequence of binary numbers may produce an error or ambiguity during transition from one number to next.
- It is used in telegraphy, robust communication, error detection and correction.

Applications of Octal Number System:

- It is used for entering the binary data and displaying certain informations.
- It is very important for the efficient use of microprocessors and other digital circuits.

Bcd To Excess - 3 Code Encoder

Excess: 3 codes is a modified form of a BCD number. It can be derived from the natural BCD code by adding 3 to each coded number.

Truth table:

B3	B2	B1	B0	E3	E2	E1	E0
0	0	0	0	0	0	1	1
0	0	0	1	0	1	0	0
0	0	1	0	0	1	0	1
0	0	1	1	0	1	1	0
0	1	0	0	0	1	1	1
0	1	0	1	1	0	0	0
0	1	1	0	1	0	0	1
0	1	1	1	1	0	1	0
1	0	0	0	1	0	1	1
1	0	0	0	1	1	0	0

K - Map Simplification

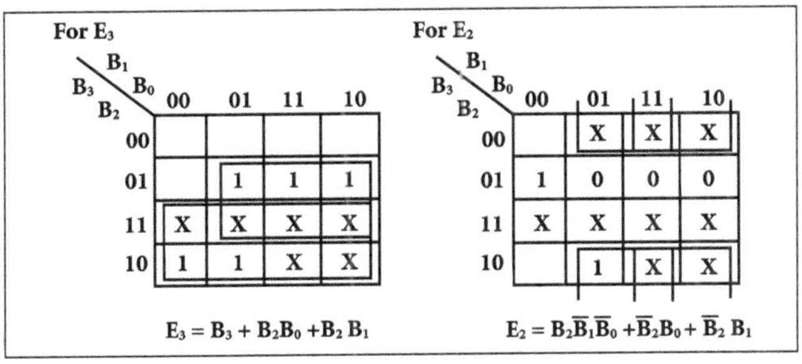

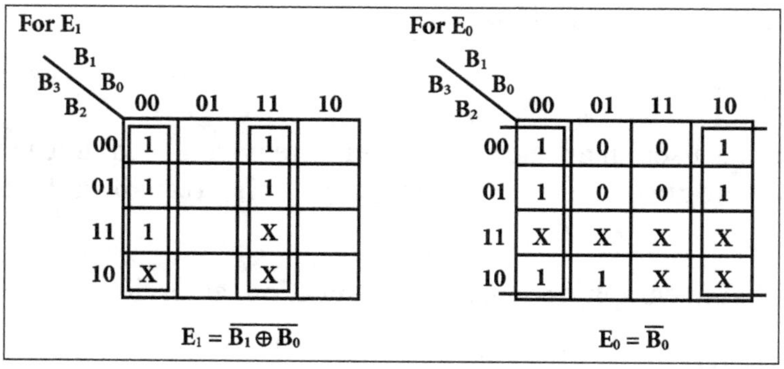

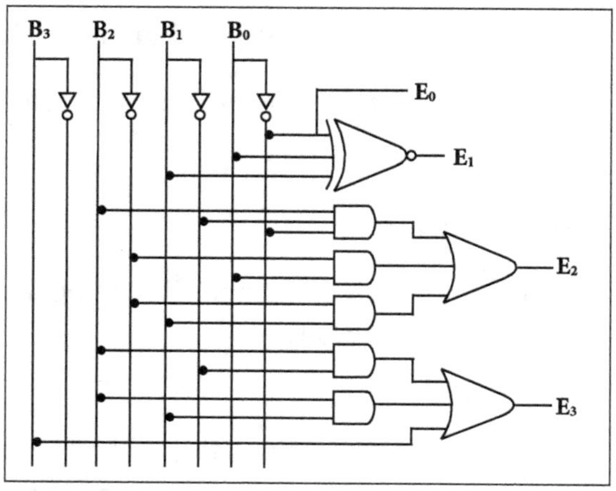

BCD to Excess 3 Code Converter.

4.3.1 Decoders

A decoder is similar to de multiplexer but without any data input. A decoder is a logic circuit which converts an n-bit binary input code (data) into 2n output lines, such that each output line shall be activated for only one of the possible combination of inputs.

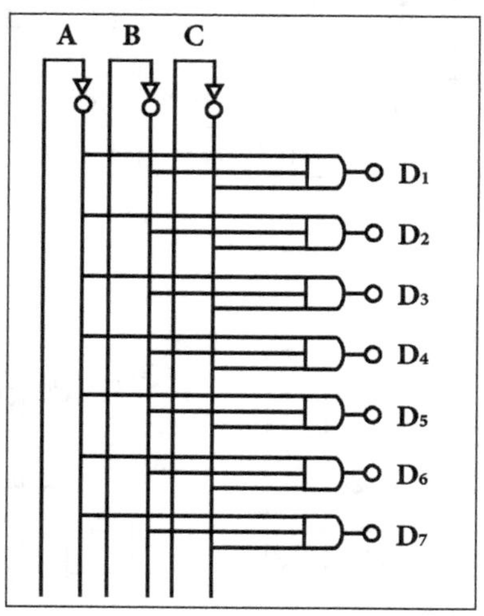

Logic Diagram of 3-8 decoder.

A 4 to 16 decoder is one with four inputs and sixteen outputs.

From the truth table given below, it is clear that, for a particular input combination, only one output is in low state while the remaining outputs are in HIGH state. The 4-to-16 decoder can be implemented by using four NOT gates and 16 decoding NAND gates.

Here, sixteen encoding gates (24 = 16) are required to decode all possible combination of four bits. Since it has four inputs and 16 outputs it is commonly called a 4 to 16 decoder. It is important to note that NAND gates are used in the implementation to have active LOW outputs.

The 4-to-16 decoder is also called a 1-of-16 decoder, since only one of the 16 outputs is selected based on a particular input combination. It is also called a binary-to-hexadecimal decoder since the inputs represent the four-bit binary number and the output represents the sixteen digits in the hexadecimal number system.

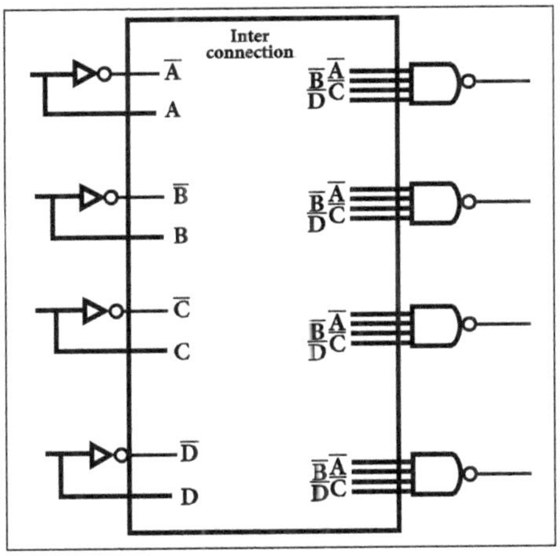

Logic diagram of 4-to-16 decoder.

3 to 8 line decoder: Three inputs are decoded into eight outputs, each output represents one of the Mini terms of the 3 input variable.

EN	A	B	C	Y_7	Y_6	Y_5	Y_4	Y_3	Y_2	Y_1	Yo
0	X	X	X	0	0	0	0	0	0	0	0
1	0	0	0	0	0	0	0	0	0	0	1
1	0	0	1	0	0	0	0	0	0	1	0
1	0	1	0	0	0	0	0	0	1	0	0
1	0	1	1	0	0	0	0	1	0	0	0
1	1	0	0	0	0	0	1	0	0	0	0
1	1	0	0	0	0	0	1	0	0	0	0

1	1	0	1	0	0	1	0	0	0	0	0
1	1	1	0	0	1	0	0	0	0	0	0
1	1	1	1	1	0	0	0	0	0	0	0

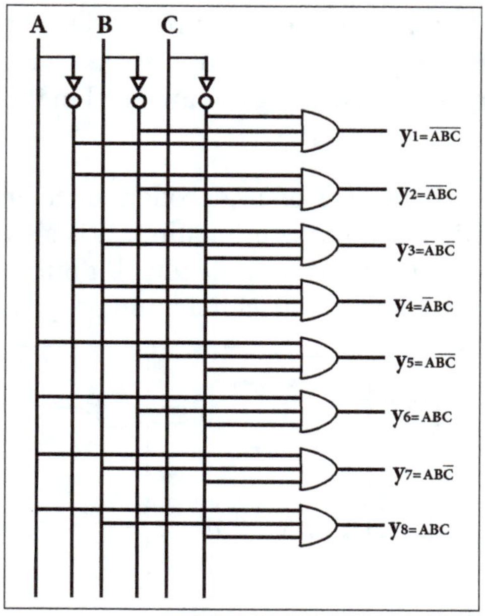

Implementation.

Segment Decoder Circuit Design

Step 1: The design involves analysis of the common cathode 7-segment display. The 7-segment display will consists of an arrangement of LEDs in an 'H' form. A truth table is constructed with the combination of inputs for each decimal number. Example, the decimal number 1 would command a combination of b and c.

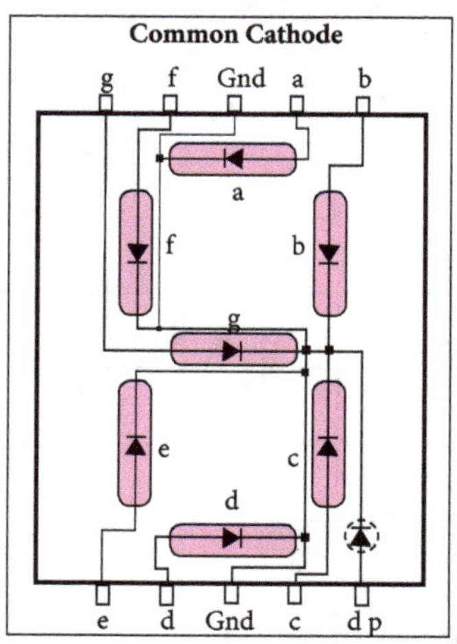

7 Segments LED.

Step 2: It involves constructing the truth table listing the 7 display input signals, decimal number and the corresponding 4 digit binary numbers.

Hence, the truth table for the decoder design depends on the type of 7-segment display. As we mentioned above, that for a common cathode seven-segment display, the output of decoder or segment driver should be active high in order to glow the segment.

The below figure shows the truth table of a BCD to seven-segment decoder with common cathode display. There are 7 different output columns corresponding to each of the 7 segments in truth table.

Let the column for segment a shows the different combinations for which it is to be illuminated. Hence, 'a' is active for the digits 0, 2, 3, 5, 6, 7, 8 and 9.

Digit	A	B	C	D	a	b	c	d	e	f	g
0	0	0	0	0	0	0	0	0	0	0	1
1	0	0	0	1	1	0	0	1	1	1	1
2	0	0	1	0	0	0	1	0	0	1	0
3	0	0	1	1	0	0	0	0	1	1	0
4	0	1	0	0	1	0	0	1	1	0	0
5	0	1	0	1	0	1	0	0	1	0	0
6	0	1	1	0	0	1	0	0	0	0	0
7	0	1	1	1	0	0	0	1	1	1	1
8	1	0	0	0	0	0	0	0	0	0	0
9	1	0	0	1	0	0	0	0	1	0	0

From the above truth table, the equation obtained are as follows:

$$a = F_1(A, B, C, D) = \sum m\ (0, 2, 3, 5, 7, 8, 9)$$

$$b = F_2(A, B, C, D) = \sum m\ (0, 1, 2, 3, 4, 7, 8, 9)$$

$$c = F_3(A, B, C, D) = \sum m\ (0, 1, 3, 4, 5, 6, 7, 8, 9)$$

$$d = F_4(A, B, C, D) = \sum m\ (0, 2, 3, 5, 6, 8)$$

$$e = F_5(A, B, C, D) = \sum m\ (0, 2, 6, 8)$$

$$f = F_6(A, B, C, D) = \sum m\ (0, 4, 5, 6, 8, 9)$$

$$g = F_7(A, B, C, D) = \sum m\ (2, 3, 4, 5, 6, 8, 9)$$

Thus, these are the Boolean expressions for each output functions.

Step 3: It involves constructing the K- map for each output term and then simplifying them for obtaining a logic combination of the inputs for each output.

K-Map Simplification

The below figures shows the k-map simplification for the common cathode seven-segment decoder in order to design the combinational circuit.

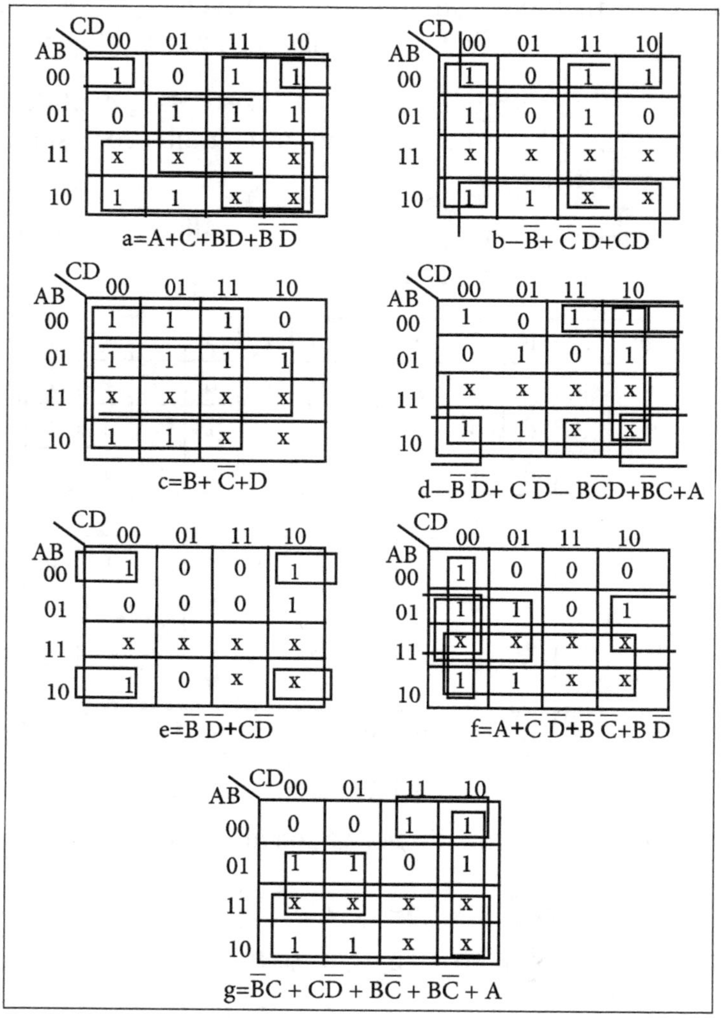

From the above simplification, we get the output values as,

$$a = A + C + BD + \overline{B}\,\overline{D}$$

$$b = \overline{B} + \overline{C}\overline{D} + CD$$

$$c = B + \overline{C} + D$$

$$d = \overline{B}\,\overline{D} + C\overline{D} + B\overline{C}D + \overline{B}C + A$$

$$e = \overline{B}\overline{D} + C\overline{D}$$

$$f = A + \overline{C}\overline{D} + B\overline{C} + B\overline{D}$$

$$g = A + B\overline{C} + \overline{B}C + C\overline{D}$$

Step 4: The final step involves drawing a combinational logic circuit for each output signal. Once the task was accomplished, a combinational logic circuit can be drawn using 4 inputs (A, B, C, D) and a 7- segment display (a, b, c, d, e, f, g) as output.

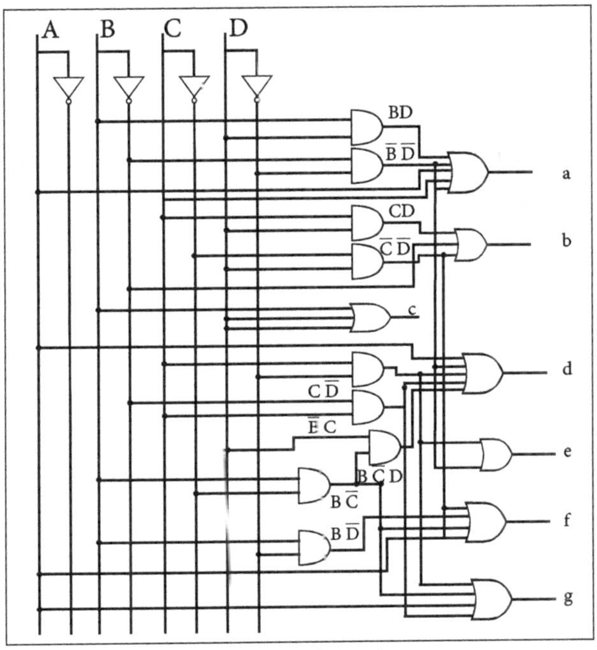

K-Map Simplification.

4.3.2 De Multiplexers

In addition to the data input there must be other inputs to control the transmission of the data to the appropriate data output line at any given time. This de multiplexer circuit having eight output lines is shown in the below figure (a).

It is instructive to compare this de multiplexer circuit with the multiplexer circuit in the below figure (b).

For the same number of control inputs, it has same number of AND gates. But now each AND gate output is a circuit output.

Rather than each gate having its own separate data input, single data line now forms one of the inputs to each AND gate, the other AND inputs being control inputs.

When the word formed by the control inputs $C_2C_1C_0$ is the binary equivalent of decimal k, then the data input x is routed to output Dk.

Otherwise, for a de multiplexer with n control inputs, each AND gate output corresponds to a mini term of n variables.

For a given combination of control inputs, only one mini term can take on the value 1, the data input is routed to the AND gate corresponding to this mini term.

For example, the logical expression for the output D_3 is $xC_2'C_1C_0$. Thus, when $C_2C_1C_0$ = 011, then D_3 = x and all other D_i are equal to 0.

The complete truth table for the eight-output de multiplexer is shown in the below figure:

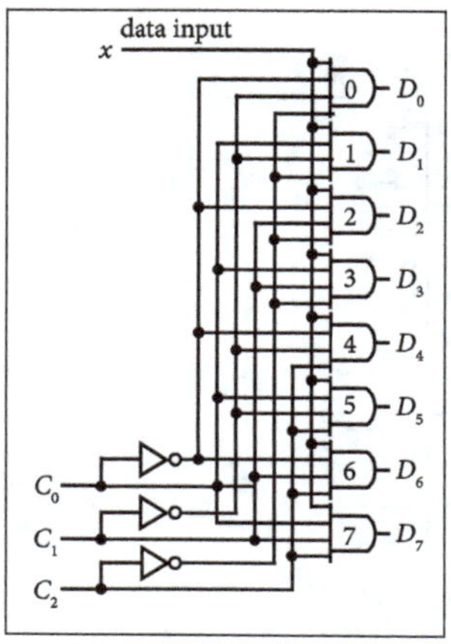

(a) De mux circuit

Control Input			Data Output							
C_2	C_1	C_0	D_0	D_1	D_2	D_3	D_4	D_5	D_6	D_7
0	0	0	x	0	0	0	0	0	0	0
0	0	1	0	x	0	0	0	0	0	0
0	1	0	0	0	x	0	0	0	0	0
0	1	1	0	0	0	x	0	0	0	0
1	0	0	0	0	0	0	x	0	0	0
1	0	1	0	0	0	0	0	x	0	0
1	1	0	0	0	0	0	0	0	x	0
1	1	1	0	0	0	0	0	0	0	x

(b) Truth table.

7 Segment Decoder Circuit Design

Step 1: The first step of the design involves analysis of the common cathode 7-segment display. A 7-segment display consists of an arrangement of LEDs in an 'H' form. A truth table is constructed with the combination of inputs for each decimal number. For example, decimal number 1 would command a combination of b and c.

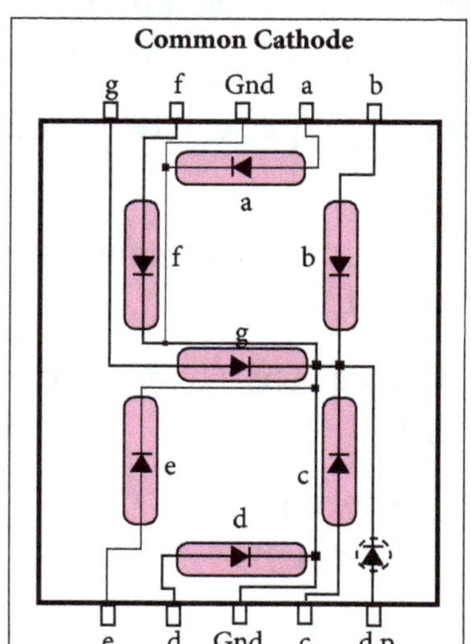

7 Segment LED.

Step 2: The second step involves constructing the truth table listing the 7 display input signals, decimal number and corresponding 4 digit binary numbers.

The truth table for the decoder design depends on the type of 7-segment display. As we mentioned above that for a common cathode seven-segment display, the output of decoder or segment driver must be active high in order to glow the segment.

The below figure shows the truth table of a BCD to seven-segment decoder with common cathode display. In the truth table, there are 7 different output columns corresponding to each of the 7 segments.

Suppose the column for segment a shows the different combinations for which it is to be illuminated. So 'a' is active for the digits 0, 2, 3, 5, 6, 7, 8 and 9.

Digit	A	B	C	D	a	b	c	d	e	f	g
0	0	0	0	0	0	0	0	0	0	0	1
1	0	0	0	1	1	0	0	1	1	1	1
2	0	0	1	0	0	0	1	0	0	1	0
3	0	0	1	1	0	0	0	0	1	1	0
4	0	1	0	0	1	0	0	1	1	0	0
5	0	1	0	1	0	1	0	0	1	0	0
6	0	1	1	0	0	1	0	0	0	0	0
7	0	1	1	1	0	0	0	1	1	1	1
8	1	0	0	0	0	0	0	0	0	0	0
9	1	0	0	1	0	0	0	0	1	0	0

From the above truth table, the Boolean expressions of each output functions can be written as,

$$a = F_1(A, B, C, D) = \sum m\ (0, 2, 3, 5, 7, 8, 9)$$

$$b = F_2(A, B, C, D) = \sum m\ (0, 1, 2, 3, 4, 7, 8, 9)$$

$$c = F_3(A, B, C, D) = \sum m\ (0, 1, 3, 4, 5, 6, 7, 8, 9)$$

$$d = F_4(A, B, C, D) = \sum m\ (0, 2, 3, 5, 6, 8)$$

$$e = F_5(A, B, C, D) = \sum m\ (0, 2, 6, 8)$$

$$f = F_6(A, B, C, D) = \sum m\ (0, 4, 5, 6, 8, 9)$$

$$g = F_7(A, B, C, D) = \sum m\ (2, 3, 4, 5, 6, 8, 9)$$

Step 3: The third step involves constructing the K-map for each output term and then simplifying them to obtain a logic combination of inputs for each output.

K-Map Simplification

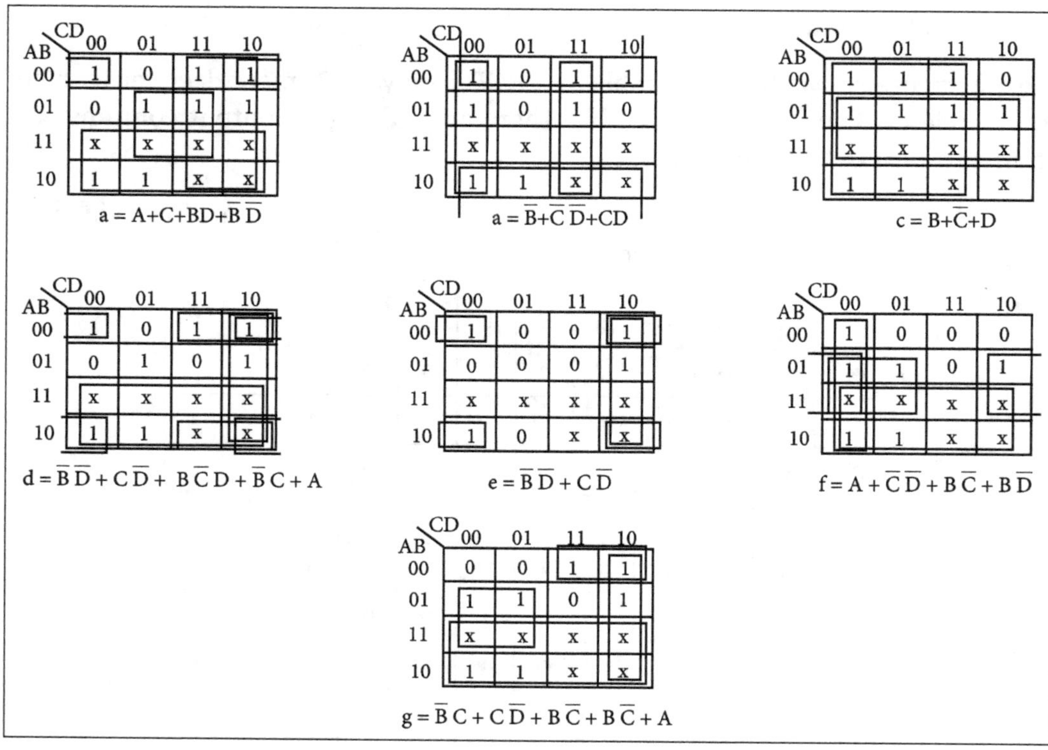

From the above simplification, we get the output values as,

$$a = A+C+BD+\overline{B}\overline{D}$$

$$b = \overline{B}+\overline{C}\overline{D}+CD$$

$$c = B+\overline{C}+D$$

$$d = \overline{B}\overline{D}+C\overline{D}+B\overline{C}D+\overline{B}C+A$$

$$e = \overline{B}\overline{D}+C\overline{D}$$

$$f = A+\overline{C}\overline{D}+B\overline{C}+B\overline{D}$$

$$g = A+B\overline{C}+\overline{B}C+C\overline{D}$$

Step 4: The final step involves drawing a combinational logic circuit for each output signal. Once the task was accomplished, a combinational logic circuit can be drawn using 4 inputs (A,B,C,D)and a 7- segment display (a, b, c, d, e, f, g) as output.

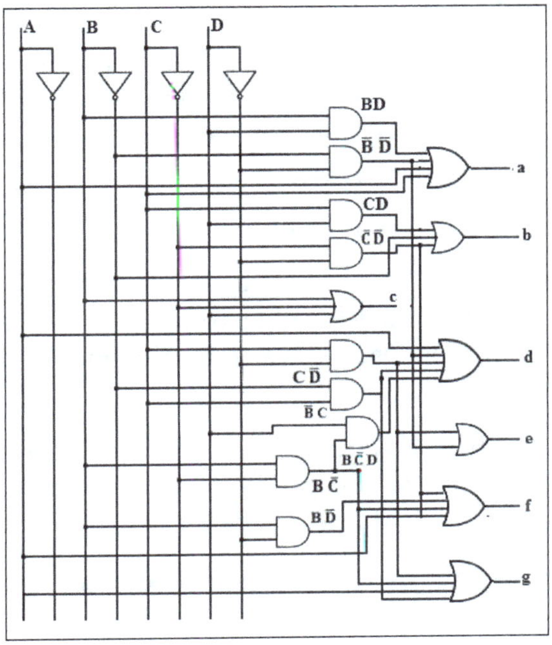

4.3.3 LED & LCD Decoders with Drivers

Display Decoder

A Digital Decoder IC, is a device which converts one digital format into another and one of the most commonly used devices for doing this is called the Binary Coded Decimal (BCD) to 7-Segment Display Decoder.

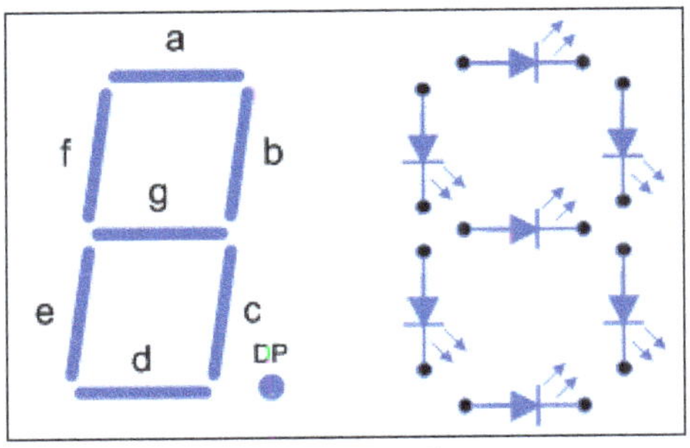

Display Deccder.

7-segment LED (Light Emitting Diode) or LCD (Liquid Crystal Display) type displays, provide a very convenient way of displaying information or digital data in the form of numbers, letters or even alpha-numerical characters.

Typically 7-segment displays consist of seven individual colored LED's (called the segments), within one single display package. In order to produce the required numbers or

HEX characters from 0 to 9 and A to F respectively, on the display the correct combination of LED segments need to be illuminated and BCD to 7-segment Display Decoders such as the 74LS47 do just those.

A standard 7-segment LED display generally has 8 input connections, one for each LED segment and one that acts as a common terminal or connection for all the internal display segments. Some single displays have also have an additional input pin to display a decimal point in their lower right or left hand corner.

In electronics there are two important types of 7-segment LED digital display.

The Common Cathode Display (CCD): In the common cathode display, all the cathode connections of the LED's are joined together to logic "0" or ground. The individual segments are illuminated by application of a "HIGH", logic "1" signal to the individual Anode terminals.

The Common Anode Display (CAD): In the common anode display, all the anode connections of the LED's are joined together to logic "1" and the individual segments are illuminated by connecting the individual Cathode terminals to a "LOW", logic "0" signal.

Common Cathode and Common Anode Format

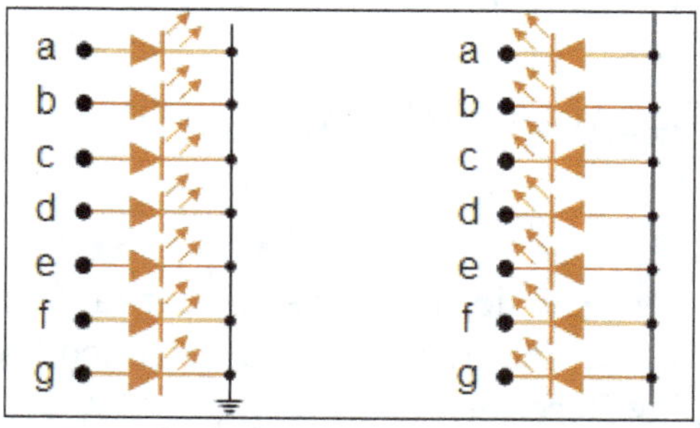

Common cathode display Common anode display.

Electrical connection of the individual diodes for a common cathode display and a common anode display and by illuminating each light emitting diode individually, they can be made to display a variety of numbers or characters.

7-Segment Display Format

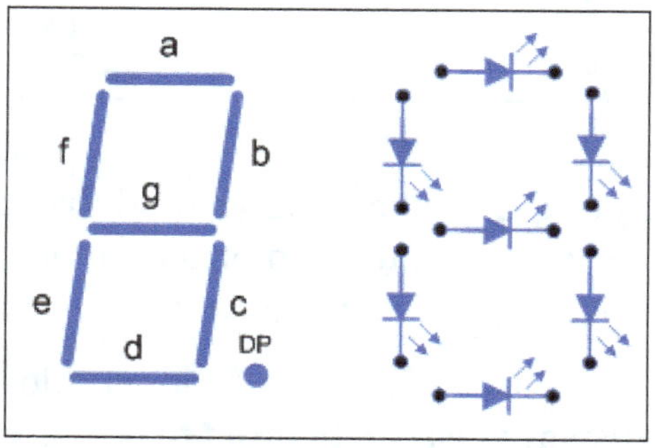

7-segment display.

So in order to display the number 3 for example, segments a, b, c, d and g would need to be illuminated. If we wanted to display a different number or letter then a different set of segments would need to be illuminated. Then for a 7-segment display, we can produce a truth table giving the segments that need to be illuminated in order to produce the required character as shown below.

Truth Table for A 7-Segment Display

7-segment display characters:

Individual Segments							Display
a	b	c	d	e	f	g	
×	×	×	×	×	×		0
	×	×					1
×	×		×	×		×	2
×	×	×	×			×	3
	×	×			×	×	4
×		×	×		×	×	5
×		×	×	×	×	×	6
×	×	×					7

Individual Segments							Display
a	b	C	d	e	f	g	
×	×	×	×	×	×	×	8
×	×	×	×		×	×	9
×	×	×		×	×	×	A
		×	×	×	×	×	b
×			×	×	×		c
	×	×	×	×		×	d
×			×	×	×	×	E
×			×	×	×	×	F

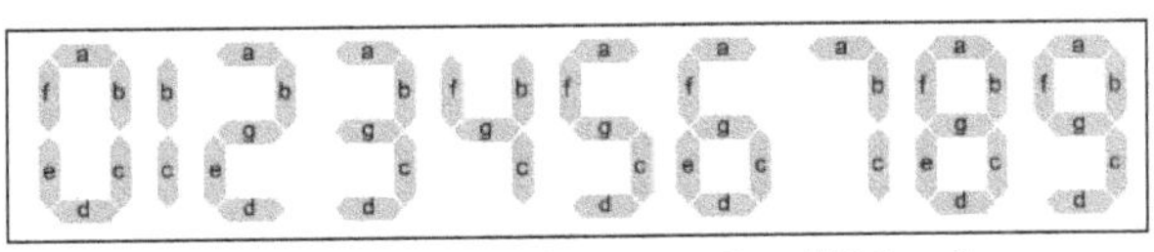

7-Segment Display Elements for all Numbers

It can be seen that to display any single digit number from 0 to 9 in binary or letters from A to F in hexadecimal, we would require 7 separate segment connections plus one additional connection for the LED's "common" connection.

Also as the segments are basically a standard light emitting diode, the driving circuit would need to produce up to 20 mA of current to illuminate each individual segment and to display the number 8, all 7 segments would need to be lit resulting a total current of nearly 140mA, (8 x 20 mA).

Obviously, the use of so many connections and power consumption is impractical for some electronic or microprocessor based circuits and so in order to reduce the number of signal lines required to drive just one single display, display decoders such as the BCD to 7-Segment Display Decoder and Driver IC's are used instead.

Binary Coded Decimal

Binary Coded Decimal (BCD or "8421" BCD) numbers are made up using just 4 data bits (a nibble or half a byte) similar to the Hexadecimal numbers, but unlike hexadecimal numbers that range in full from 0 through to F, BCD numbers only range from 0 to 9, with the binary number patternss of 1010 through to 1111 (A to F) being invalid inputs for this type of display and so are not used as shown below.

Decimal	Binary Pattern				BCD
	8	4	2	1	
0	0	0	0	0	0
1	0	0	0	1	1
2	0	0	1	0	2
3	0	0	1	1	3
4	0	1	0	0	4
5	0	1	0	1	5
6	0	1	1	0	6
7	0	1	1	1	7

Decimal	Binary Pattern				BCD
	8	4	2	1	
8	1	0	0	0	8
9	1	0	0	0	9
10	1	0	1	1	Invalid
11	1	0	1	1	Invalid
12	1	1	0	0	Invalid
13	1	1	0	0	Invalid
14	1	1	1	1	Invalid
15	1	1	1	1	Invalid

BCD to 7-Segment Display Decoders

A binary coded decimal (BCD) to 7-segment display decoder such as the TTL 74LS47 or 74LS48, have 4 BCD inputs and 7 output lines, one for each LED segment. This allows a smaller 4-bit binary number (half a byte) to be used to display all the denary numbers from 0 to 9 and by adding two displays together a full range of numbers from 00 to 99 can be displayed with just a single byte of 8 data bits.

BCD to 7-Segment Decoder

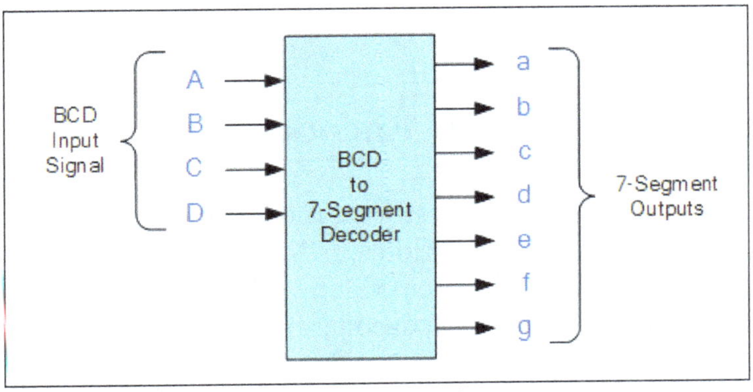

BCD display decoder.

The use of packed BCD allows two BCD digits to be stored within a single byte (8-bits) of data, allowing a single data byte to hold a BCD number in the range of 00 to 99.

An example of the 4-bit BCD input (0100) representing the number 4 is given below.

Example 1:

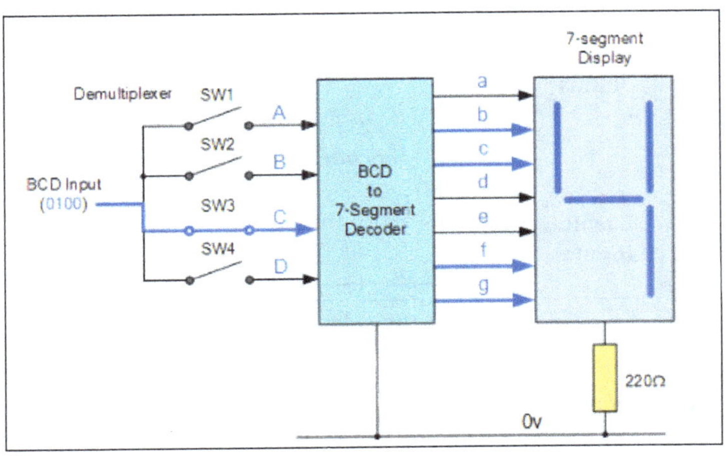

In practice current limiting resistors of about 150Ω to 220Ω would be connected in series between the decoder/driver chip and each LED display segment to limit the maximum current flow. There are different display decoders and drivers available for the different types of available displays, either LED or LCD. For example, the 74LS48 for common-cathode LED types, the 74LS47 for common-anode LED types, or the CMOS CD4543 for liquid crystal display (LCD) types.

Liquid crystal displays (LCD´s) have one major advantage over similar LED types in that they consume much less power and nowadays, both LCD and LED displays are combined together to form larger Dot-Matrix Alphanumeric type displays which can show letters and characters as well as numbers in standard Red or Tri-colour outputs.

4.4 Encoders and Priority Encoders

An encoder is the digital circuit that performs the inverse operation of a decoder. An encoder has 2n input lines and n output lines. In the encoder, the output lines generate the binary code corresponding to the input value. An encoder is a combinational circuit that performs the inverse operation of a decoder.

When the device output code has fewer bits than the input code has, the device is usually called an encoder. e.g., 2n-to-n priority encoders. It is a 2^n-to-n binary encoder, where it has only 2n inputs and the output is the n-bit binary number corresponding to the active input.

The below figure shows the general structure of the encoder circuit. As shown in the below figure, the decoded information is presented as 2n inputs producing n possible outputs.

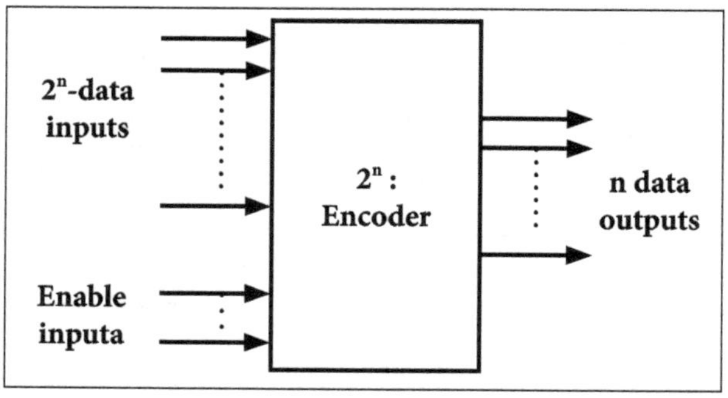

Encoder

4.4.1 Priority Encoder

A priority encoder is an encoder circuit that includes the priority function. In priority encoder, if two or more inputs are equal to 1 at the same time, the input having the highest priority will take precedence.

The below table shows truth table of 4-bit priority encoder.

Input				Output		
D_0	D_1	D_2	D_3	Y_1	Y_0	V
0	0	0	0	X	X	0
1	0	0	0	0	0	1
X	1	0	0	0	1	1
X	X	1	0	1	0	1
X	X	X	1	1	1	1

Truth table of 4-bit priority encoder.

The above table shows D_3 input with highest priority and D_0 input with lowest priority. When D_3 input is high, regardless of other inputs output is 11. The D_2 has the next priority. Thus, when $D_3 = 0$ and $D_2 = 1$, regardless of other two lower priority input, output is 10.

The output for D1 is generated only if higher priority inputs are 0 and so on. The output V (a valid output indicator) indicates, one or more of the inputs are equal to 1. If all inputs are 0, V is equal to 0 and the other two outputs (Y_1 and Y_0) of the circuit are not used.

K-Map Simplification

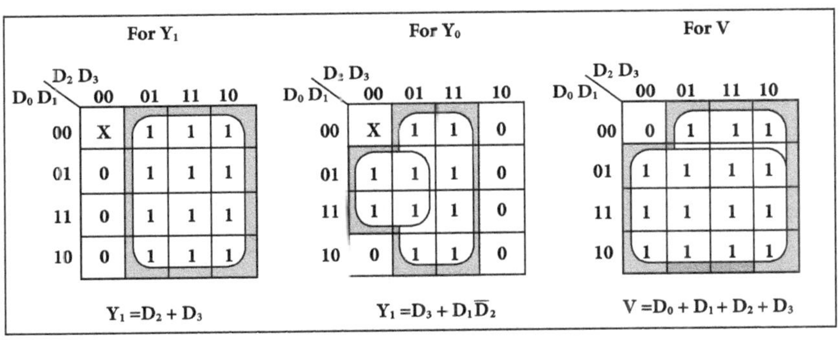

Logic Diagram

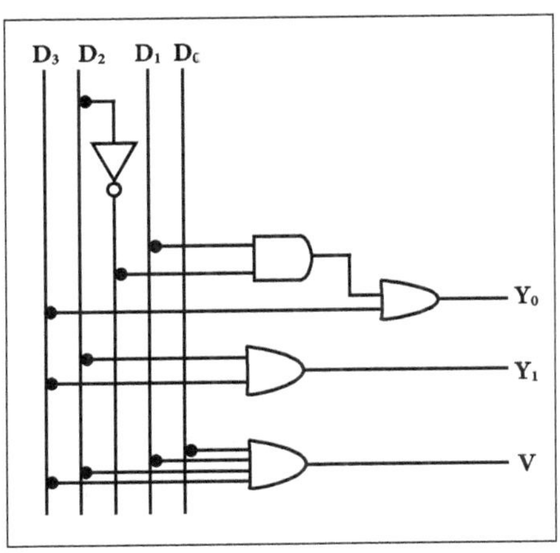

4.5 Multiplexers, DE-Multiplexers and Priority Generators/Checkers

Multiplexer

Multiplexer is a digital switch. It allows digital information from several sources to be routed onto a single output line. Multiplexer is a special type of combinational circuit. It has n data inputs, one output and select inputs with $2^m = n$.

Depending on the digital code applied at the select inputs, one out of a data sources is selected and transmitted to the single output Y. E is called as a strobe or enable input which is useful for cascading. It is generally an active low terminal that means it will perform required operation when it is low.

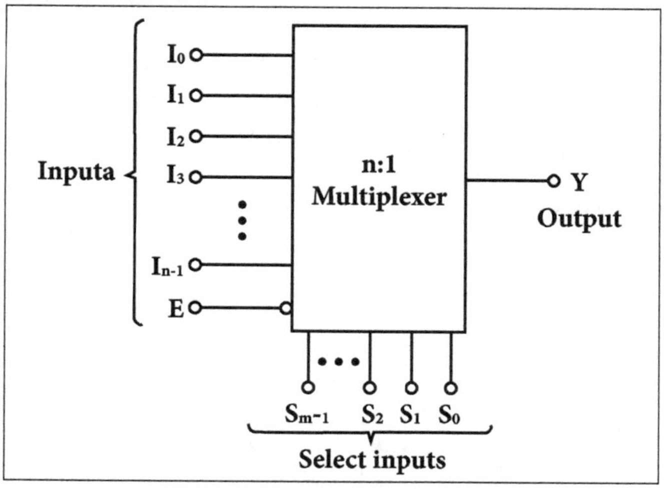

Data routing.

Multiplexers can be used to route the data from one of several sources to one destination.

Control Sequences

A Multiplexer can also be used as a part of control sequences.

2 X 2 Multiplexer

Multiplication of binary number is performed in the same way as in decimal numbers. The multiplicand is multiplied by each bit of the multiplier which starts from the least significant bit. Each such multiplication forms a partial product. Successive partial products are shifted one position to the left. The final product is obtained from the sum of partial products.

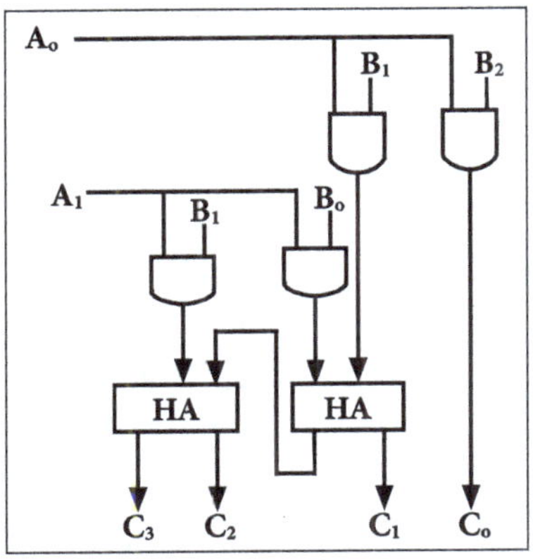

2 x 2 multiplexer.

The multiplication of 2 bit number is considered. The multiplicand bits are B1 and B0, the multiplier bits are A1 and A0 and the product is $C_3 C_2 C_1 C_0$. The first partial product is formed by multiplying A0 by B_1B_0.

The multiplication of two bits such as A0 and B0 produces 1 if both bits are 1, otherwise produces 0. This is identical to AND Operation. Therefore, the partial product can be implemented with AND gates as,

$$\begin{array}{cccc} & & B_1 & B_0 \\ & & A_1 & A_0 \\ \hline & & A_0B_1 & A_0B_0 \\ & A_1B_1 & A_1B_0 & \\ \hline C_3 & C_2 & C_1 & C_0 \end{array}$$

The second partial product is formed by multiplying A1 by B1B0 and shifted one position to left. The two partial products are added with two half adder circuits. The LSB of the product does not have to go through an adder since it is formed by the output of the first AND gate.

Multiplexer Types

Multiplexers are classified as:

8-to-1 Multiplexer

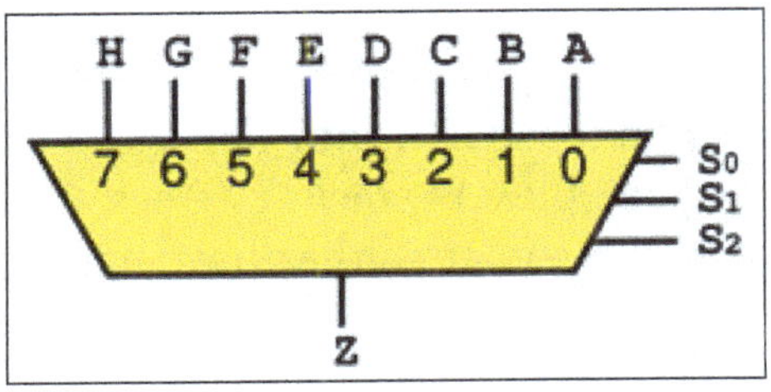

8-to-1 Multiplexer.

The 8-to-1 multiplexer consists of 8 input lines, one output line and 3 selection lines.

8-1 Multiplexer Circuit

For the combination of selection input, the data line is connected to the output line. The circuit shown below is an 8*1 multiplexer. The 8-to-1 multiplexer requires 8 AND gates, one OR gate and 3 selection lines. As an input, the combination of selection inputs are giving to the AND gate with the corresponding input data lines.

In a similar fashion, all the AND gates are given connection. In this 8*1 multiplexer, for any selection line input, one AND gate gives a value of 1 and the remaining all AND gates give 0. And finally, by using OR gate, all the AND gates are added; and, this will be equal to the selected value.

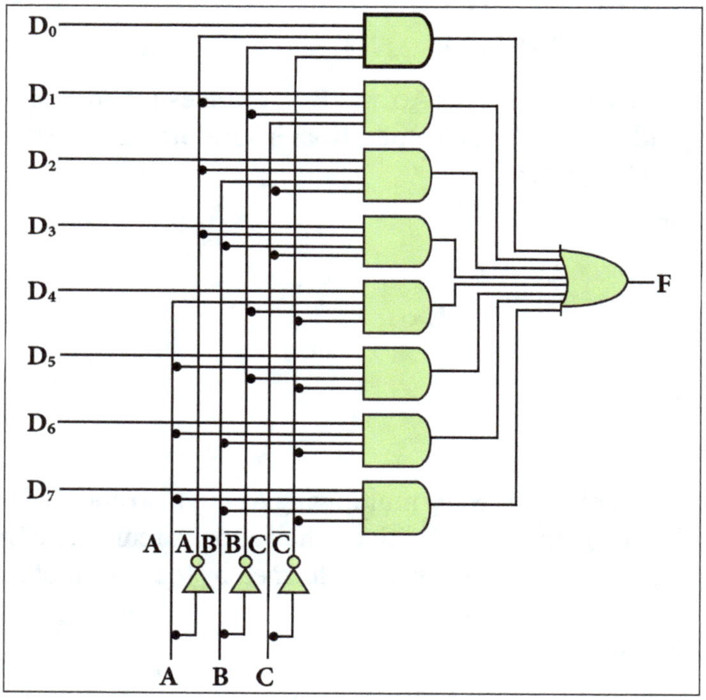

8-1 Multiplexer Circuit.

Higher Order Multiplexing

The 16-to-1 multiplexers are the largest available ICs. Hence, to meet the larger input needs there must be a provision for expansion. This can be achieved with the help of enable or strobe inputs and multi-plexers stacks or trees are designed. Two commonly used methods for this purpose are shown in the above figure (a) and (b).

The circuit of below figure (a) uses two 16:1 multiplexers (M_1 and M_2) for the realization of a 32: 1 multiplexer.

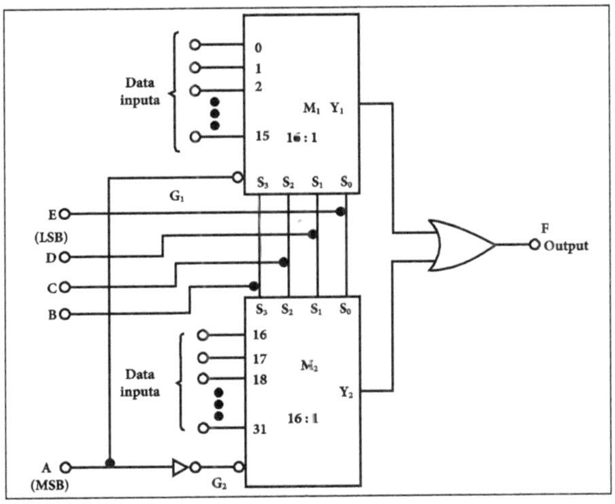

(a) 16 : 1 mux with OR gate.

The lower order of 16 data input lines (I_0 - I_{15}) are applied at the data input terminals of the multiplexer M^1 and the higher order 16 data input lines (I_{15} - I_{31}) are applied at the data input terminals of the multiplexer M_2.

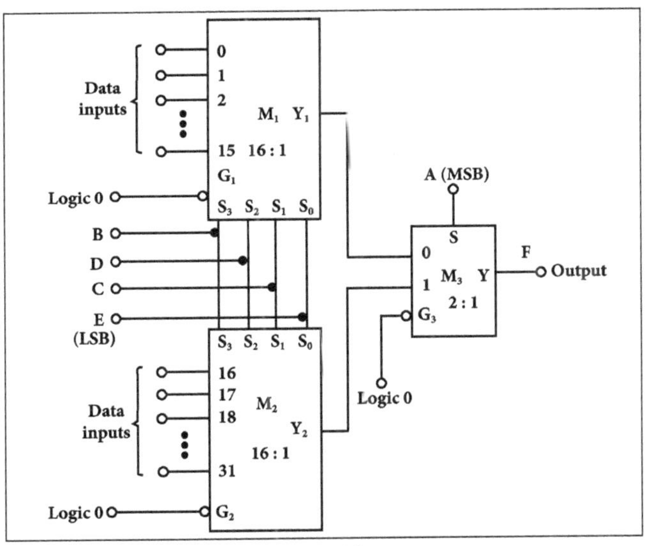

(b) 16 : 1 mux with 2 :1 mux.

For a 32:1 multiplexer, the number of select input lines is required to be 5. The 5-bit select inputs are ABCDE. The most-significant select input bit A is applied at G_1 and $\bar{A}$ is applied at G_2. B, C, D and E are connected to S_3, S_2, S_1 and S_0 inputs of both the multiplexers. The output F of the multiplexer is obtained by using an OR gate, where $F = Y_1 + Y_2$.

When A = 0, the multiplexer M_1 is enabled and M_2 is disabled, thereby allowing one of the lower order 16 bits to Y_1, depending upon the value of BCDE. $Y_2 = 0$. Therefore, $F = Y_1$. Similarly, when A = 1, M_2 is enabled and M_1 is disabled and $F = Y_2$.

The above figure (b) uses a 2:1 multiplexer instead of an OR gate. When A = 0, F = Y_1 and when A = 1, F = Y_2. Thus, both the circuits perform similar operation. These two general techniques can be used to expand to an n input multiplexer without any difficulty.

Applications of Multiplexers: Multiplexers are used in various applications wherein multiple-data need to be transmitted by using single line.

Communication System: A communication system has both a communication network and a transmission system. By using a multiplexer, the efficiency of the communication system can be increased by allowing the transmission of data, such as audio and video data from different channels through single lines or cables.

Telephone Network: In telephone networks, multiple audio signals are integrated on a single line of transmission with the help of a multiplexer.

Computer Memory: Multiplexers are used in computer memory to maintain a huge amount of memory in the computers, and also to reduce the number of copper lines required to connect the memory to other parts of the computer.

Transmission from the Computer System of a Satellite: Multiplexer is used to transmit the data signals from the computer system of a spacecraft or a satellite to the ground system by using a GSM satellite.

4.5.1 De Multiplexers

In addition to the data input there must be other inputs to control the transmission of the data to the appropriate data output line at any given time.

Types of De Multiplexer

De-multiplexers are classified into four types:

- 1-2 de multiplexer (1 select line).
- 1-4 de multiplexer (2 select lines).
- 1-8 de multiplexer (3 select lines).
- 1-16 de multiplexer (4 select lines).

1-2 De Multiplexer (1 Select Line)

A 1-to-2 de multiplexer consists of one input line, two output lines and one select line. The signal on the select line helps to switch the input to one of the two outputs. The figure below shows the block diagram of a 1-to-2 de multiplexer with additional enable input.

In the figure, there are only two possible ways to connect the input to output lines, thus only one select signal is enough to do the de multiplexing operation. When the select input is low, then the input will be passed to Y_0 and if the select input is high then the input will be passed to Y_1.

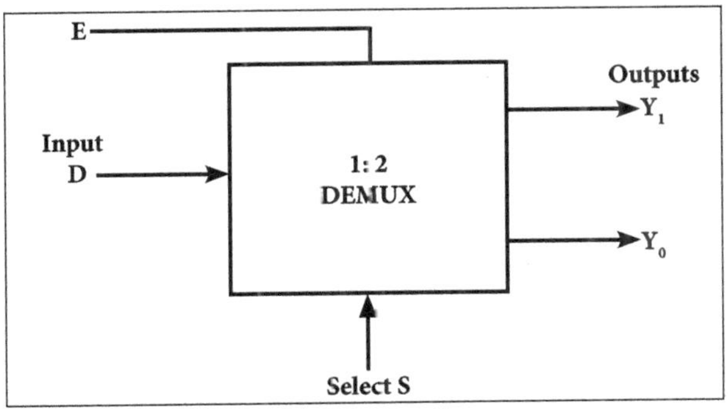

The truth table of a 1-to-2 de multiplexer is shown below in which the input is routed to Y_0 and Y_1 depends on the value of select input S. In the table output Y_1 is active when the combination of select line and input line are active high, i.e., SF = 11.

Therefore, the output Y_1 = SF and similarly the output Y_0 is equal to S_F.

Select	Input	Output	
S	D	Y_1	Y_0
0	0	0	0
0	1	0	1
1	0	0	0
1	1	1	0

From the above truth table, the logic diagram of this de multiplexer can be designed by using two AND gates and one NOT gate as shown in below figure. When the select lines S=0, AND gate A_1 is enabled while A_2 is disabled.

Then, the data from the input flows to the output line Y_1. Similarly, when S=1, AND gate A_2 is enabled and AND gate A_1 is disabled, thus data is passed to the Y_0 output.

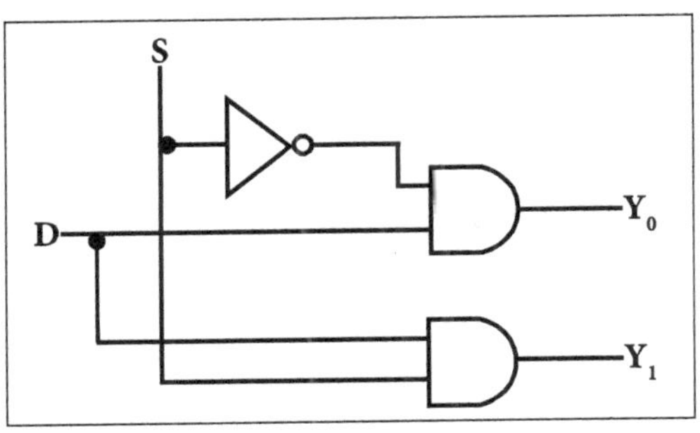

1-4 De Multiplexer (2 Select Lines)

A 1-to-4 de multiplexer has a single input (D), two selection lines (S_1 and S_0) and four outputs (Y_0 to Y_3). The input data goes to any one of the four outputs at a given time for a particular combination of select lines.

This de multiplexer is also called as a 2-to-4 de multiplexer which means that two select lines and 4 output lines. The block diagram of 1:4 DEMUX is shown below:

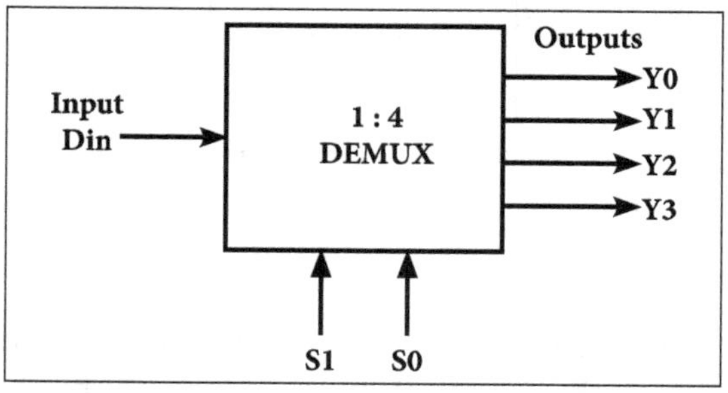

The truth table of this type of de multiplexer is given below. From the truth table it is clear that, when S_1=0 and S_0= 0, the data input is connected to output Y_0 and when S_1= 0 and S_0=1, then the data input is connected to output Y_1.

Similarly, other outputs are connected to the input for other two combinations of select lines.

Data Input	Select Inputs		Outputs			
D	S_1	S_0	Y_3	Y_2	Y_1	Y_0
D	0	0	0	0	0	D
D	0	1	0	0	D	0
D	1	0	0	D	0	0
D	1	1	D	0	0	0

From the table, the output logic can be expressed as min terms and are given below:

$$Y0 = \overline{S1}\ \overline{S0}\ D$$

$$Y1 = \overline{S1}\ S0\ D$$

$$Y2 = S1\ \overline{S0}\ D$$

$$Y3 = S1\ S0\ D$$

Where D is the input data, Y_0 to Y_3 are output lines and S_0 & S_1 are select lines.

From the above Boolean expressions, a 1-to-4 de multiplexer can be implemented by using four 3-input AND gates and two NOT gates as shown in figure below. The two selection lines enable the particular gate at a time.

So depends on the combination of select inputs, input data is passed through the selected gate to the associated output.

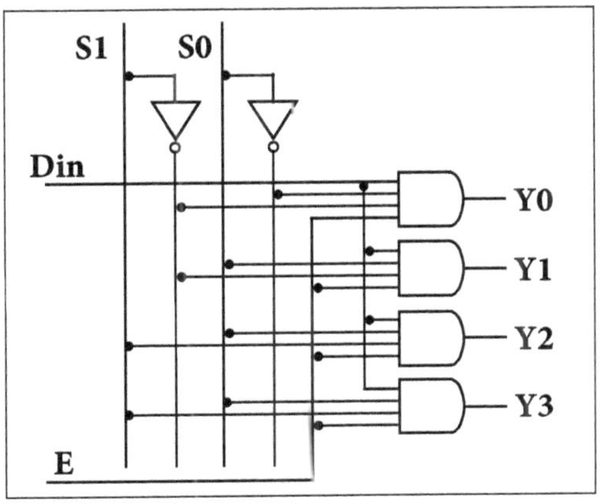

This type of de multiplexer is available in IC form and a typical IC 74139 is most commonly used dual 1-to-4 de multiplexer. It has two independent de multiplexers and each DEMUX accepts two binary inputs as select lines and four mutually exclusive active-low outputs.

Both de multiplexers share a common set of selection lines so they are selected in parallel. Also, each de multiplexer consists of enable pin or data input, for one de multiplexer it is active high data input and for other it is active low data input.

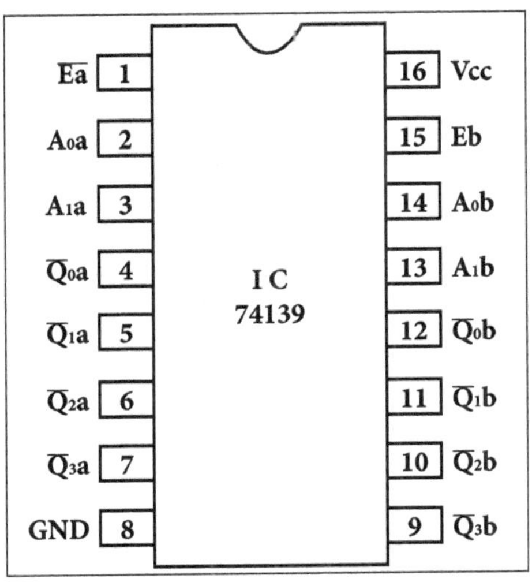

1-8 de multiplexer (3 select lines).

The below figure shows the block diagram of a 1-to-8 de multiplexer that consists of single input D, three select inputs S_2, S_1 and So and eight outputs from Y_0 to Y_7.

It is also called as 3-to-8 de multiplexer due to three select input lines. It distributes one input line to one of 8 output lines depending on the combination of select inputs.

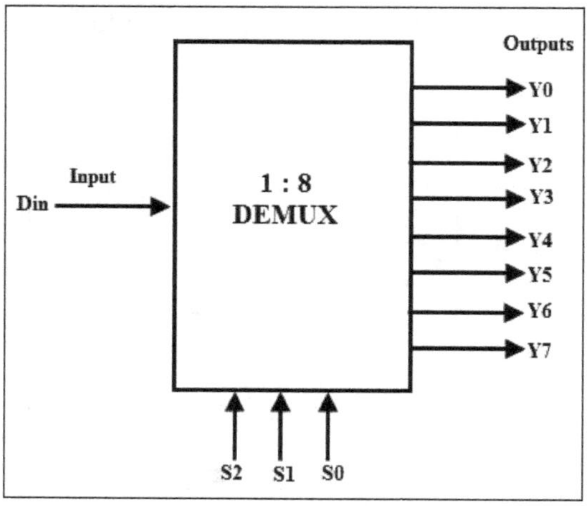

The truth table for this type of de multiplexer is shown below. The input D is connected with one of the eight outputs from Y_0 to Y_7 based on the select lines S_2, S_1 and So.

For example, if S_2 S_1 So = 000, then the input D is connected to the output Yo and so on.

Data Input	Select Input			Outputs							
D	S_2	S_1	S_0	Y_7	Y_6	Y_5	Y_4	Y_3	Y_2	Y_1	Y_0
D	0	0	0	0	0	0	0	0	0	0	D
D	0	0	1	0	0	0	0	0	0	D	0
D	0	1	0	0	0	0	0	0	D	0	0
D	0	1	1	0	0	0	0	D	0	0	0
D	1	0	0	0	0	0	D	0	0	0	0
D	1	0	1	0	0	D	0	0	0	0	0
D	1	1	0	0	D	0	0	0	0	0	0
D	1	1	1	D	0	0	0	0	0	0	0

From this truth table, the Boolean expressions for all the outputs can be written as follows,

$$Yo = D\overline{S2}\ \overline{S1}\ \overline{So}$$

$$Y1 = D\overline{S2}\ \overline{S1}\ So$$

$$Y2 = D\overline{S2}\ S1\ \overline{So}$$

$$Y3 = D\overline{S2}\,S1\,\overline{So}$$

$$Y4 = D\,S2\,\overline{S1}\,\overline{So}$$

$$Y5 = D\,S2\,\overline{S1}\,So$$

$$Y6 = D\,S2\,S1\,\overline{So}$$

$$Y7 = D\,S2\,S1\,So$$

From these obtained equations, the logic diagram of this de multiplexer can be implemented by using eight AND gates and three NOT gates as shown in below figure. The different combinations of the select lines select one AND gate at given time, such that data input will appear at a particular output.

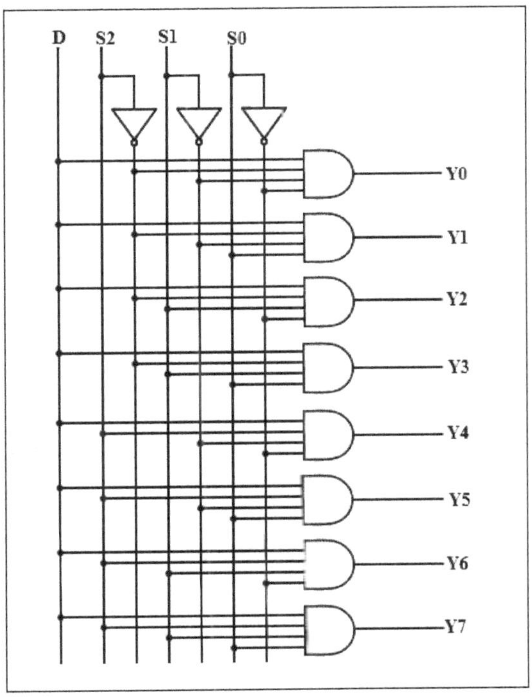

A typical IC74237 is a 1-to-8 de multiplexer that consists of latches at three select inputs. The pin out of this IC is given below.

The pins A_0 to A_2 are data inputs, Y_0 to Y_7 are de multiplexer outputs, E_1&E_2 are active-low data enable and active-high data enable pins respectively, LE is the latch enable input,Vcc and GND terminals are positive supply voltage and ground terminals.

With a 3-bit storage latch, this IC combines the 3-to-8 decoder function.

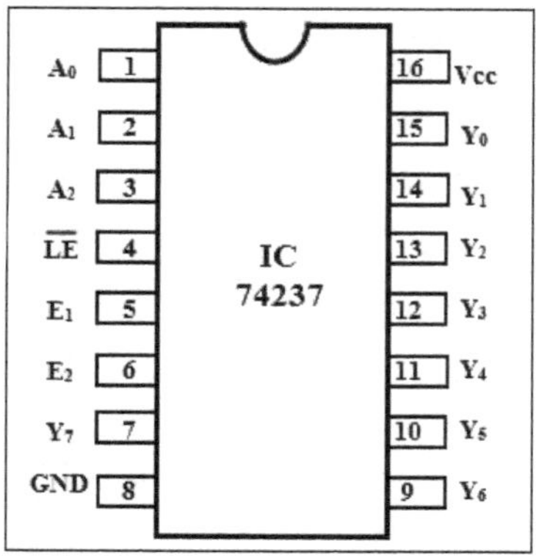

1-16 De Multiplexer (4 Select Lines)

1 to 16 de multiplexer has one input data, four select lines A, B, C and D and 16 output lines Y0 to Y15. This is implemented using AND and NOT gate. 1 to 16 de multiplexer is implemented using the logic circuit below:

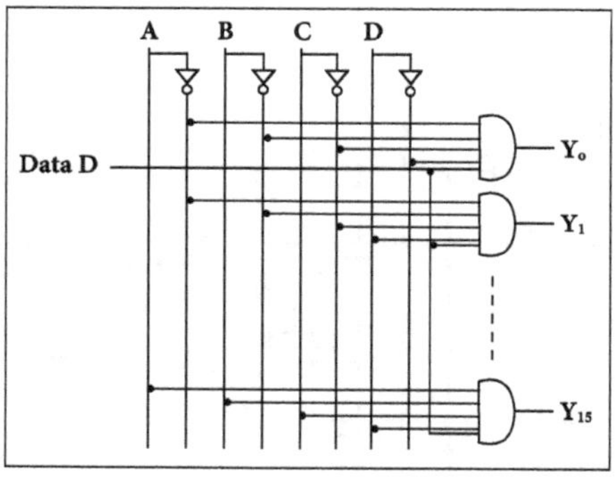

This can be implemented using 1 to 8 de multiplexer, 1 to 4 de multiplexer and 1 to 2 de multiplexer.

Truth Table

The below truth table shows the operation of 1 to 16 de multiplexer:

INPUTS						OUTPUTS															
G1	G2	D	C	B	A	0	1	2	3	4	5	6	7	8	9	10	11	12	13	14	15
L	L	L	L	L	L	L	H	H	H	H	H	H	H	H	H	H	H	H	H	H	H
L	L	L	L	L	H	H	L	H	H	H	H	H	H	H	H	H	H	H	H	H	H
L	L	L	L	H	L	H	H	L	H	H	H	H	H	H	H	H	H	H	H	H	H
L	L	L	L	H	H	H	H	H	L	H	H	H	H	H	H	H	H	H	H	H	H
L	L	L	H	L	L	H	H	H	H	L	H	H	H	H	H	H	H	H	H	H	H
L	L	L	H	L	H	H	H	H	H	H	L	H	H	H	H	H	H	H	H	H	H
L	L	L	H	H	L	H	H	H	H	H	H	L	H	H	H	H	H	H	H	H	H

L	L	L	H	H	H	H	H	H	H	H	H	H	L	H	H	H	H	H	H	H	H
L	L	H	L	L	L	H	H	H	H	H	H	H	H	L	H	H	H	H	H	H	H
L	L	H	L	L	H	H	H	H	H	H	H	H	H	H	L	H	H	H	H	H	H
L	L	H	L	H	L	H	H	H	H	H	H	H	H	H	H	L	H	H	H	H	H
L	L	H	L	H	H	H	H	H	H	H	H	H	H	H	H	H	L	H	H	H	H
L	L	H	H	L	L	H	H	H	H	H	H	H	H	H	H	H	H	L	H	H	H
L	L	H	H	L	H	H	H	H	H	H	H	H	H	H	H	H	H	H	L	H	H
L	L	H	H	H	L	H	H	H	H	H	H	H	H	H	H	H	H	H	H	L	H
L	L	H	H	H	H	H	H	H	H	H	H	H	H	H	H	H	H	H	H	H	L
L	H	X	X	X	X	H	H	H	H	H	H	H	H	H	H	H	H	H	H	H	H
H	L	X	X	X	X	H	H	H	H	H	H	H	H	H	H	H	H	H	H	H	H
H	H	X	X	X	X	H	H	H	H	H	H	H	H	H	H	H	H	H	H	H	H
H = High level,									L = Low level,								X = if relevant				

Higher Order De Multiplexers

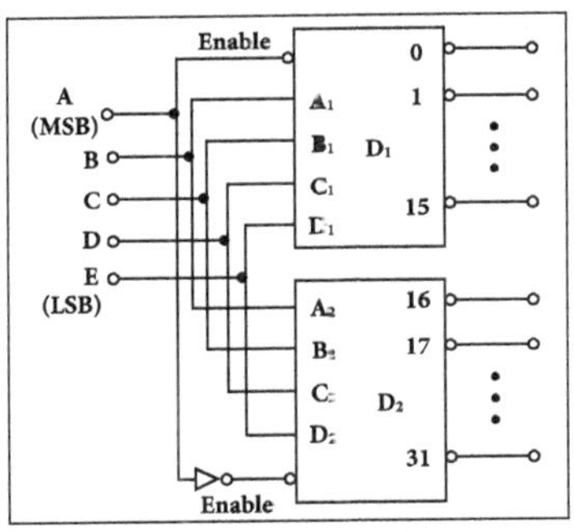

(a) 5-line-to 32-line decoder using two 4-line-to 16-line decoders.

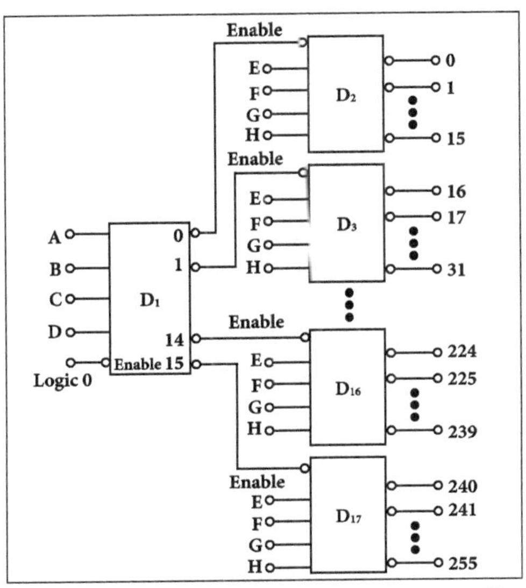

(b) 8-line-to 256-line decoder using two 4-line-to 16-line decoders.

Since 4-line-to 16-line decoders are the largest available circuits in 1Cs, to meet the larger inputs need there should be a provision for expansion. This is made possible by using enable input terminal. The above figure (a) shows a 5-line-to 32-line decoder and the above figure (b) shows a 8-line-to 256-line decoder using 4-line-to 16-line decoders.

Similarly, any m-line-to-n-line decoder can be implemented. However, if only a few codes of a large number need be recognized, the alternative approach Higher Order Multiplexing (16:1 Mux with OR gate) can be used. This is connected to detect the digital number 00011111.

The most-significant 4-bits are applied at ABCD inputs and the least-significant bits are applied at EFGH inputs. The output goes low when the most-significant bits are 0001 and the least-significant bits are 1111. The circuit can be expanded to detect other 8-bit codes.

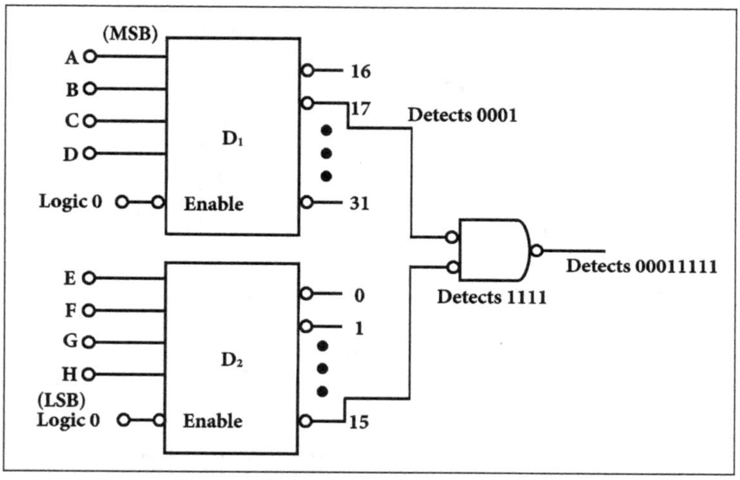

(c) An alternative approach to decode combinations.

Applications of De Multiplexer

De multiplexers are used to connect a single source to multiple destinations. These applications include the following:

Serial to Parallel Converter: This converter is used to reconstruct parallel data. In this technique, serial data is given as an input to the De-multiplexer at a regular interval, and a counter is attached to the de multiplexer at the control input to detect the data signal at the output of the de multiplexer. When all data signals are stored, the output of the de mux can be read out in parallel.

Communication System: Mux and de mux both are used in communication system to carry out the process of data transmission. A De-multiplexer receives the output signals from the multiplexer and at the receiver end it converts them back to the original form.

Arithmetic Logic Unit: The output of the ALU is fed as an input to the De-multiplexer and the output of the de multiplexer is connected to a multiple register. The output of the ALU can be stored in multiple registers.

4.5.2 Priority Generators/Checkers

Priority Generator

It is combinational circuit that accepts an n-1 bit stream data and generates the additional bit that is to be transmitted with the bit stream. This additional or extra bit is termed as a parity bit.

In even parity bit scheme, the parity bit is '0' if there are even number of 1s in the data stream and the parity bit is '1' if there are odd number of 1s in the data stream.

In odd parity bit scheme, the parity bit is '1' if there are even number of 1s in the data stream and the parity bit is '0' if there are odd number of 1s in the data stream.

Even Parity Generator

Let us assume that a 3-bit message is to be transmitted with an even parity bit. Let the three inputs A, B and C are applied to the circuits and output bit is the parity bit P. The total number of 1s must be even, to generate the even parity bit P.

The below figure shows the truth table of even parity generator in which 1 is placed as parity bit in order to make all 1s as even when the number of 1s in the truth table is odd.

Even Parity Generator Truth table:

3 – bit message			Even parity bit generator (P)
A	B	C	Y
0	0	0	0
0	0	1	1
0	1	0	1
0	1	1	0
1	0	0	1
1	0	1	0
1	1	0	0
1	1	1	1

The K-map simplification for 3-bit message even parity generator is,

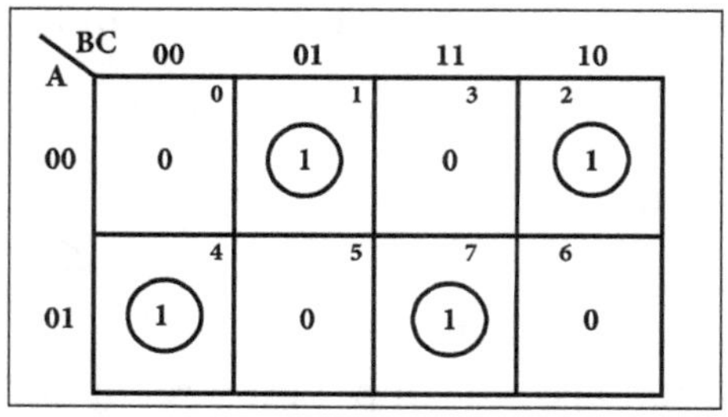

K map for Even Parity Generator.

From the above truth table, the simplified expression of the parity bit can be written as,

$$P=\bar{A}\,\bar{B}\,C+\bar{A}\,B\,\bar{C}+A\,\bar{B}\,\bar{C}+A\,B\,C$$

$$=\bar{A}\left(\bar{B}C+B\bar{C}\right)+A\left(\bar{B}\,\bar{C}+BC\right)$$

$$=\bar{A}\left(B\oplus C\right)+A\left(\overline{B\oplus C}\right)$$

$$P=A\oplus B\oplus C$$

The above expression can be implemented by using two Ex-OR gates. The logic diagram of even parity generator with two Ex – OR gates is shown below. The three bit message along with the parity generated by this circuit which is transmitted to the receiving end where parity checker circuit checks whether any error is present or not.

To generate the even parity bit for a 4-bit data, three Ex-OR gates are required to add the 4-bits and their sum will be the parity bit.

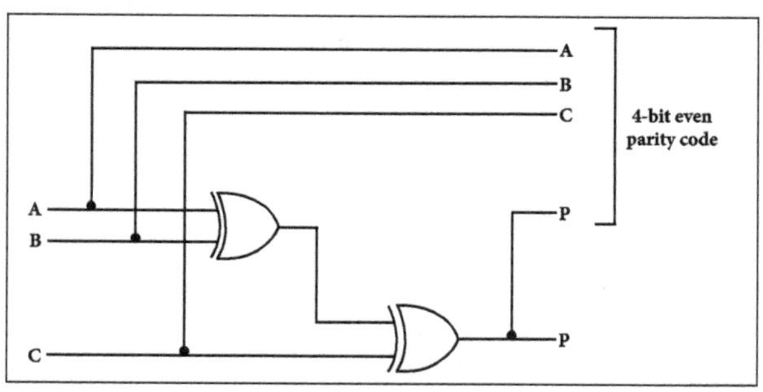

Even Parity Generator Logic Circuit.

Odd Parity Generator

Let us consider that the 3-bit data is to be transmitted with an odd parity bit. The three inputs are A, B and C and P is the output parity bit. The total number of bits must be odd in order to generate the odd parity bit.

In the given truth table below, 1 is placed in the parity bit in order to make the total number of bits odd when the total number of 1s in the truth table is even.

Odd Parity Generator Truth table:

3 – bit message			Even parity bit generator (P)
A	B	C	Y
0	0	0	1
0	0	1	0
0	1	0	0
0	1	1	1
1	0	0	0
1	0	1	1
1	1	0	1
1	1	1	0

The truth table of the odd parity generator can be simplified by using K-map as,

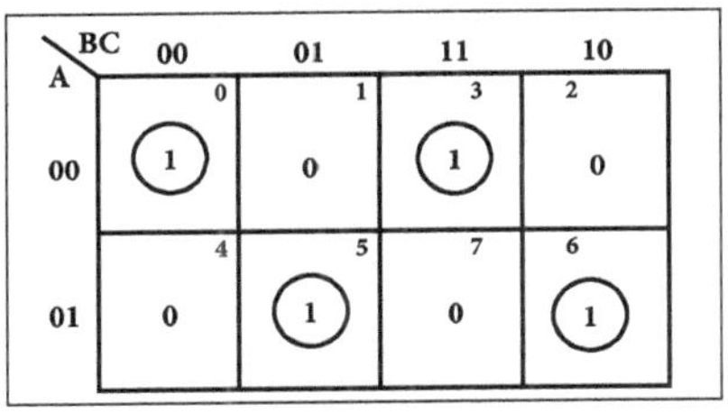

K map for Odd Parity Generator.

The output parity bit expression for this generator circuit is obtained as,

$$P = A \oplus B \text{ Ex} - \text{NOR C}$$

The above Boolean expression can be implemented by using one Ex-OR gate and one Ex-NOR gate in order to design a 3-bit odd parity generator.

The logic circuit of this generator is shown in below figure, in which two inputs are applied at one Ex-OR gate, and this Ex-OR output and third input is applied to the Ex-NOR gate, to produce the odd parity bit. It is also possible to design this circuit by using two Ex-OR gates and one NOT gate.

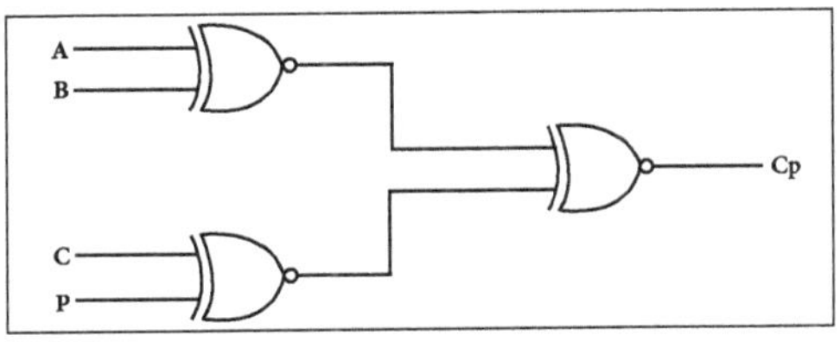

Odd Parity Checker Logic Circuit.

Parity Check

It is a logic circuit that checks for possible errors in the transmission. This circuit can be an even parity checker or odd parity checker depending on the type of parity generated at the transmission end. When this circuit is used as even parity checker, the number of input bits must always be even.

When a parity error occurs, the 'sum even' output goes low and 'sum odd' output goes high. If this logic circuit is used as an odd parity checker, the number of input bits should be odd, but if an error occurs the 'sum odd' output goes low and 'sum even' output goes high.

Even Parity Checker

Consider that three input message along with even parity bit is generated at the transmitting end. These 4 bits are applied as input to the parity checker circuit which checks the possibility of error on the data. Since the data is transmitted with even parity, four bits received at circuit must have an even number of 1s.

If any error occurs, the received message consists of odd number of 1s. The output of the parity checker is denoted by PEC (parity error check).

The below table shows the truth table for the even parity checker in which PEC = 1 if the error occurs, i.e., the four bits received have odd number of 1s and PEC = 0 if no error occurs, i.e., if the 4-bit message has even number of 1s.

Even Parity Checker Truth table:

4- bit received message				Parity error check C_p
A	B	C	P	
0	0	0	0	0
0	0	0	1	1
0	0	1	0	1
0	0	1	1	0
0	1	0	0	1
0	1	0	1	0
0	1	1	0	0
0	1	1	1	1
1	0	0	0	1
1	0	0	1	0
1	0	1	0	0
1	0	1	1	1
1	1	0	0	0
1	1	0	1	1
1	1	1	0	0
1	1	1	1	0

The above truth table can be simplified using K-map as shown below:

AB \ CP	00	01	11	10
00	0 (0)	1 (1)	0 (3)	1 (2)
01	1 (4)	0 (5)	1 (7)	0 (6)
11	0 (12)	1 (13)	0 (15)	1 (14)
10	1 (8)	0 (9)	1 (11)	0 (10)

K map for Even Parity Checker,

$$PEC=\bar{A}\,\bar{B}\left(\bar{C}D+C\bar{D}\right)+\bar{A}B\left(\bar{C}\bar{D}+CD\right)+AB\left(\bar{C}D+C\bar{D}\right)+A\bar{B}\left(\bar{C}\bar{D}+CD\right)$$

$$=\bar{A}\,\bar{B}\,(C\oplus D)\bar{A}B\left(\overline{C\oplus D}\right)+AB(C\oplus D)+A\bar{B}\left(\overline{C\oplus D}\right)$$

$$=\left(\bar{A}\,\bar{B}+AB\right)(C\oplus D)+\left(\bar{A}B+A\bar{B}\right)\left(\overline{C\oplus D}\right)$$

$$=(A\oplus B)\oplus(C\oplus D)$$

The above logic expression for the even parity checker can be implemented by using three Ex-OR gates as shown in the below figure. If the received message consists of five bits, then one more Ex-OR gate is required for the even parity checking.

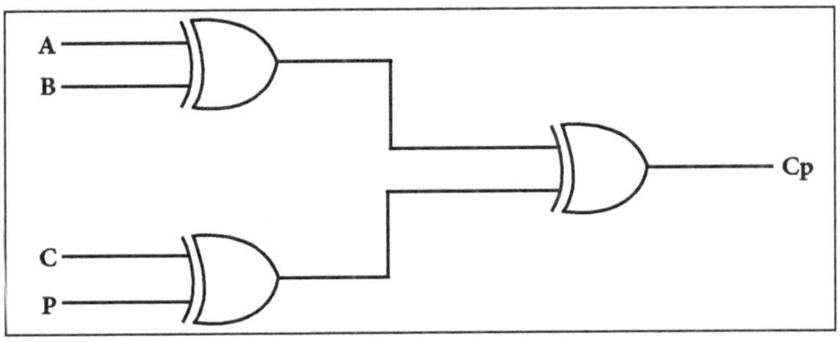

Even Parity Checker Logic Circuit.

Odd Parity Checker

Consider that a three bit message along with odd parity bit is transmitted at the transmitting end. Odd parity checker circuit receives these 4 bits and checks whether any error are present in the data.

If the total number of 1s in the data is odd, then it indicates no error, whereas if the total number of 1s is even then it indicates the error since the data is transmitted with odd parity at transmitting end.

The below figure shows the truth table for odd parity generator where PEC = 1 if the 4-bit message received consists of even number of 1s (hence the error occurred) and PEC = 0 if the message contains odd number of 1s (that means no error).

Odd Parity Checker Truth table:

4- bit received message				Parity error check C_p
A	B	C	P	
0	0	0	0	1
0	0	0	1	0
0	0	1	0	0
0	0	1	1	1
0	1	0	0	0
0	1	0	1	1
0	1	1	0	1
0	1	1	1	0
1	0	0	0	0
1	0	0	1	1
1	0	1	0	1
1	0	1	1	0
1	1	0	0	1
1	1	0	1	0
1	1	1	0	0
1	1	1	1	1

The expression for the PEC in the above truth table can be simplified by K-map as shown below:

AB \ CP	00	01	11	10
00	0: 1	1: 0	3: 1	2: 0
01	4: 0	5: 1	7: 0	6: 1
11	12: 1	13: 0	15: 1	14: 0
10	8: 0	9: 1	11: 0	10: 1

K map for Odd Parity Checker.

After simplification, the final expression for the PEC is obtained as,

$$PEC = (A\ Ex-NOR\ B) Ex-NOR (C\ Ex-NOR\ D)$$

The expression for the odd parity checker can be designed by using three Ex-NOR gates as shown below:

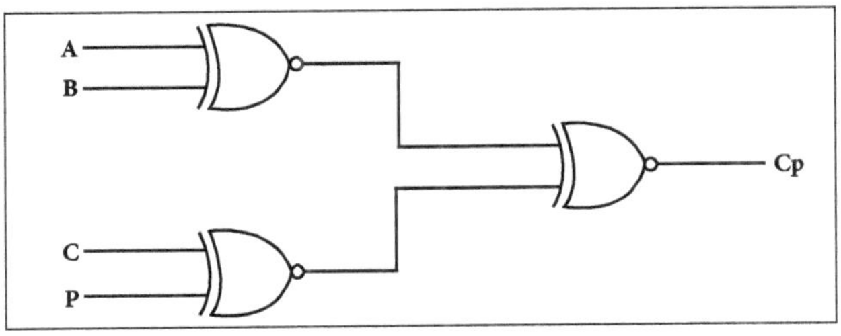

Odd Parity Checker Logic Circuit.

Parity Generator/Checker ICs

There are different types of parity generator /checker ICs are available with different input configurations such as 5-bit, 4-bit, 9-bit, 12-bit, etc. A most commonly used and standard type of parity generator/checker IC is 74180.

It is a 9-bit parity generator or checker used to detect errors in high speed data transmission or data retrieval systems. The figure below shows the pin diagram of 74180 IC.

This IC can be used to generate a 9-bit odd or even parity code or it can be used to check for odd or even parity in a 9-bit code (8 data bits and one parity bit).

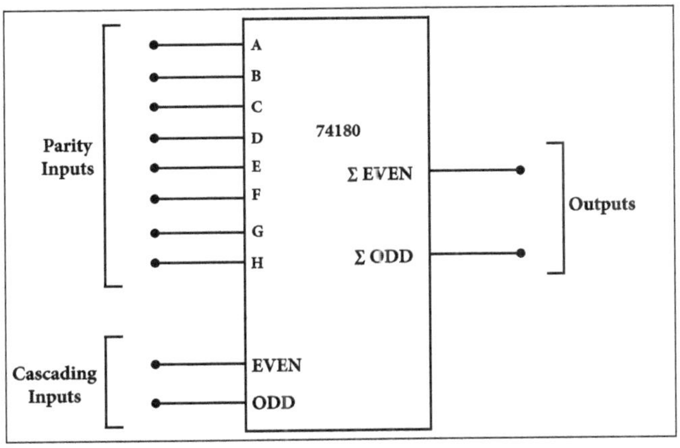

IC 74180.

This IC consists of eight parity inputs from A through H and two cascading inputs. There are two outputs even sum and odd sum. In implementing generator or checker circuits, unused parity bits must be tied to logic zero and the cascading inputs must not be equal.

4.6 Parallel Binary Adder and Subtract: Magnitude Comparators

Binary Parallel Adder (IC 74LS83/74LS283)

Many high-speed adders available in integrated-circuit form utilize the look ahead carry or a similar technique for reducing overall propagation delays. The most common is a 4-bit parallel adder IC (741S83/74283) that contains four inter connected full-adders and the look ahead carry circuitry needed for high-speed operation. The 7483 and 74283 are a Tit medium scale integrated (MSI) circuit with same pin configuration.

The below figure shows the logic diagram and the functional symbol for the 74LS283 4-bit parallel adder. The inputs to this IC are two 4-bit numbers, A_3, A_2, A_1, A_0 and B_3, B_2, B_1, B_0 and the carry, Cino, into the LSB position. The outputs are the sum bits S_3, S_2, S_1, S_0, and the carry, Cout3, output of the MSB position.

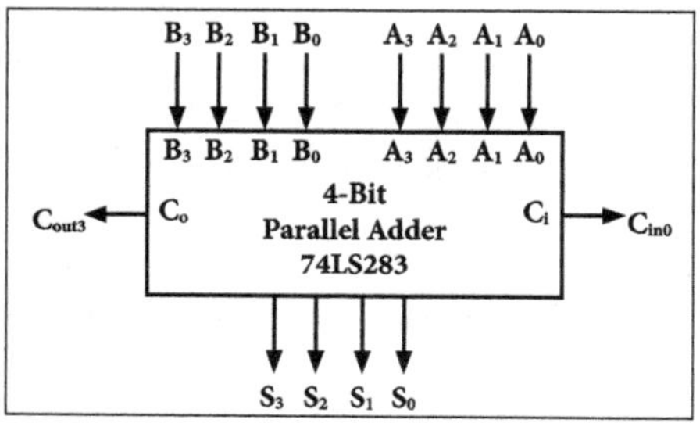

(a) Functional symbol for the 74LS283.

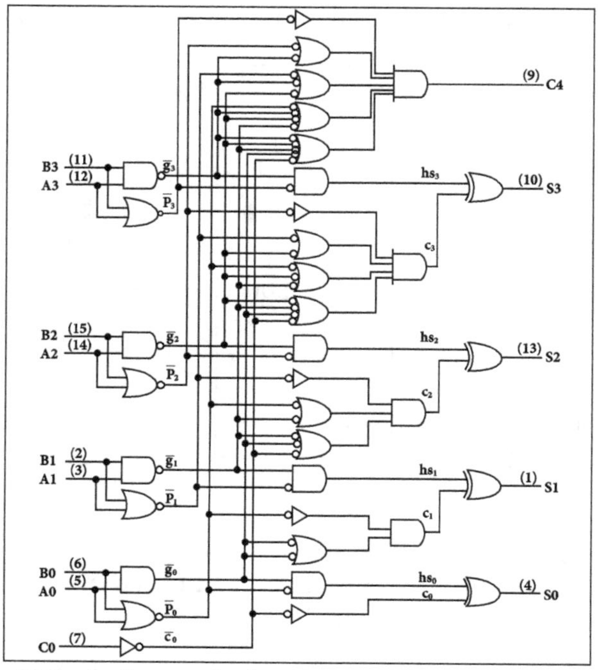

(b) Logic diagram for the 74LS283 4-bit binary adder.

Iere, the C_{i+1} equation uses the term P_i. G_i, instead of G_i. This has no effect on the output, ince Pi is always 1 when Gi is 1. However, it always the equation to be factorized as follows,

$$C_{i+1} = P_i \cdot G_i + P_i \cdot C_i$$
$$= P_i \left(G_i + C_i\right)$$

This gives us the following carry equations, which are used by the circuit of 74LS283.

$$C_1 = P_0 \cdot (G_0 + C_0)$$
$$C_2 = P_1 \cdot (G_1 + C_1)$$
$$= P_1 \cdot (G_1 + P_0 \cdot (G_0 + C_0))$$
$$= P_1 \cdot (G_1 + P_0) \cdot (G_1 + G_0 + C_0)$$
$$C_3 = P_2 \cdot (G_2 + C_2)$$
$$= P_2 \cdot (G_2 + P_1 \cdot (G_1 + P_0) \cdot (G_1 + G_0 + C_0))$$
$$= P_2 \cdot (G_2 + P_1) \cdot (G_2 + G_1 + P_0) \cdot (G_2 + G_1 + G_0 + C_0)$$
$$C_4 = P_3 \cdot (G_3 + C_3)$$
$$= P_3 \cdot (G_3 + P_2 \cdot (G_2 + P_1) \cdot (G_2 + G_1 + P_0) \cdot (G_2 + G_1 + G_0 + C_0))$$
$$= P_3 \cdot (G_3 + P_2) \cdot (G_3 + G_2 + P_1) \cdot (G_3 + G_2 + G_1 + P_0) \cdot\cdot (G_3 + G_2 + G_1 + G_0 + C_0)$$

Two or more parallel-adder blocks can be connected (cascaded) to accommodate the addition of larger binary numbers.

Let us take it as example design an 8-bit adder using two 74283s:

Solution:

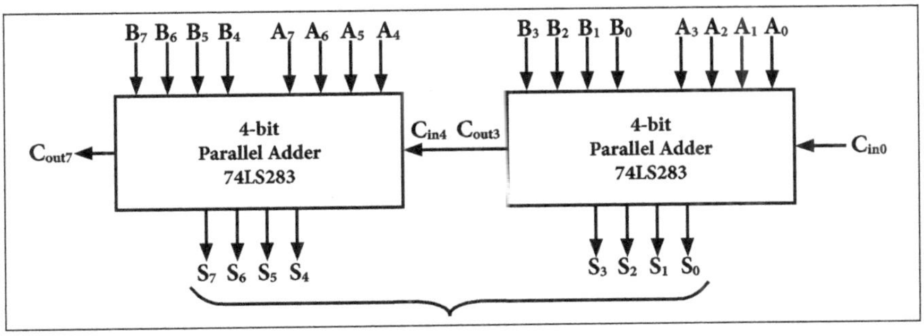

The above figure shows how two 7413283 adders can be connected to add two 8-bit numbers. The adder on the right adds the four least significant bits of the numbers. The Cout3 output of this adder is connected as the input carry to the first position of the second adder, which adds the four must significant bits of the numbers.

The eight sum outputs represent the resultant sum of the two 8-bit numbers. $C_{out}7$ is th carry out of the last position (MSB) of the second adder. It can be used as an overflo bit or as a carry into another adder stage if large binary numbers are to be handled.

Let us take it as example design a 24-bit group ripple adder using 74X283 ICs:

Solution:

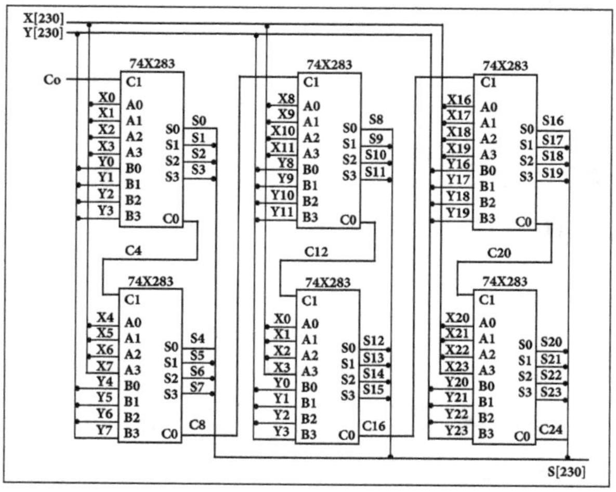

A 24-bit group ripple adder using 74X283 ICs.

Parallel Binary Sub Tractor

N-Bit Parallel Sub Tractor

The subtraction of binary numbers can be done most conveniently by means of complements. Remember that the subtraction A-B can be done by taking the 2's complement of B and adding it to A. The 2's complement can be obtained by taking the 1's complement and adding one to the least significant pair of bits. The 1's complement can be implemented with inverters and a one can be added to the sum through the input carry, as shown in the below figure:

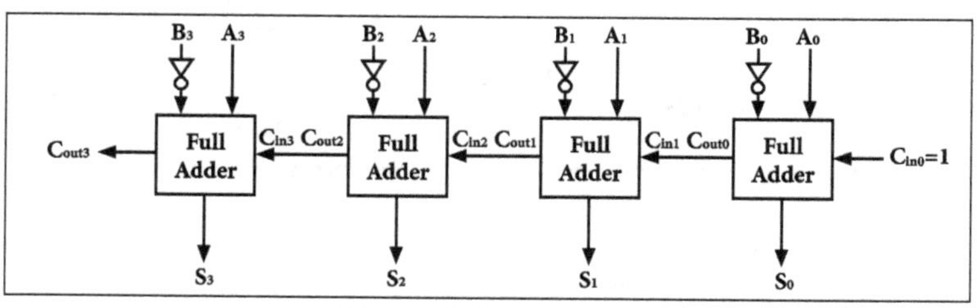

4-bit parallel sub tractor.

Binary Adder-Sub Tractor

The addition and subtraction operations can be combined into one circuit with one common binary adder. This is done by including an exclusive OR gate with each full dder, as shown in the below figure. The mode input M controls the operation of the ircuit. When M = 0, the circuit is an adder, and when M = 1, the circuit becomes a sub ractor. Each exclusive OR gate receives input M and one of the inputs of B.

Vhen M = 0, we have $B \oplus 0 = B$. The full-adders receive the value of B, the input carry s 0, and the circuit performs A plus B. When M = 1, we have $B \oplus 1 = \overline{B}$ and $C_{in\,0} = 1$. The 3 inputs are all complemented and a 1 is added through the input carry. The circuit erforms the operation A plus the 2's complement of B, i.e. A - B.

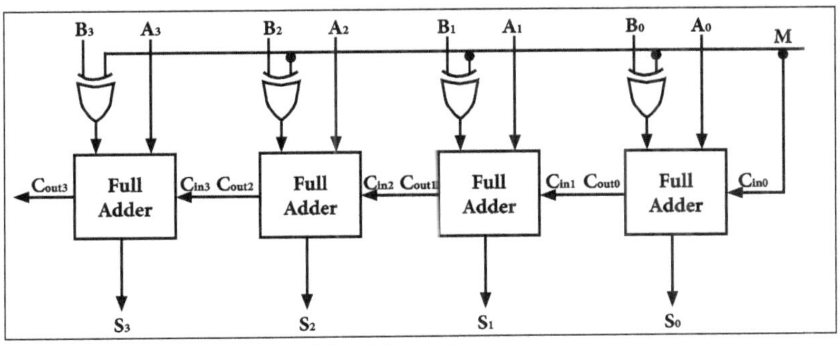

4-bit adder sub tractor.

4.6.1 Magnitude Comparators

Comparators

Digital comparator is a combinational circuit, designed to compare the two n-bit binary words applied at its input.

Comparator has three outputs A > B, A = B and A < B depending on the result of comparison of one of their outputs.

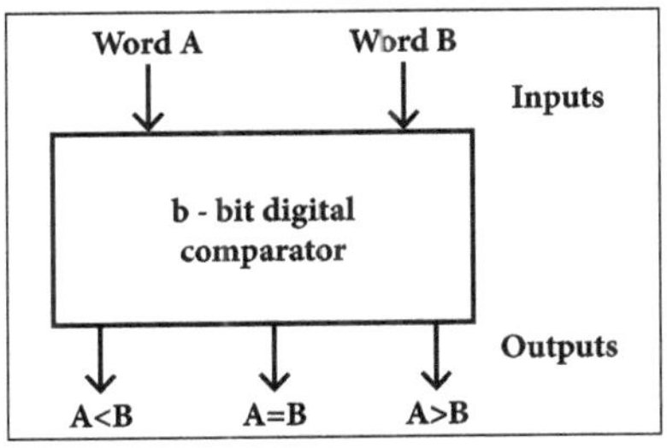

Comparator.

Magnitude Comparator

It is a combinational circuit that compares to numbers and determines their relative magnitude.

The output of comparator is usually 3 binary variables indicating:

A>B

A=B

A<B

The magnitude comparator can also be used to indicate equality, but has a further two outputs, one that is logic 1 when word A is greater than word B, and another that is logic 1 when word A is less than word B. Magnitude comparators therefore form the basis of decision making in logic circuits.

Any logical problem can be reduced to one or more (sometimes many) yes/no decisions based on a pair of compared values.

A simple 1-bit magnitude comparator is shown in the below figure. Gate 1 produces the function A > B and gate 3 gives A < B while gate 2 is an XNOR gate giving an equality output.

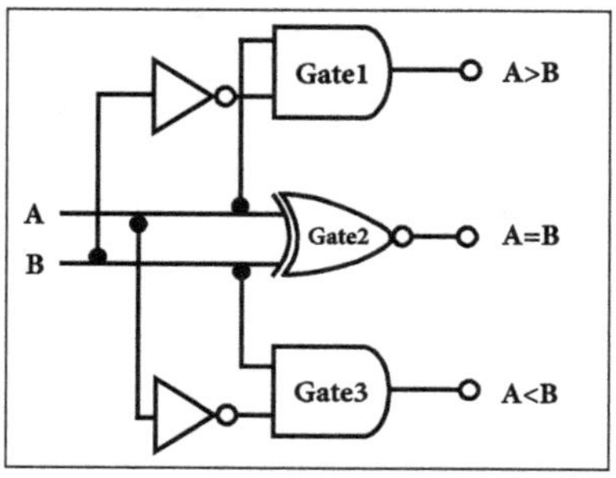

One bit magnitude comparator.

This basic circuit for a magnitude comparator may be extended for any number of bits but the more bits the circuit has to compare, the more complex the circuit becomes. Integrated circuit magnitude comparators are available that can be used to provide comparisons between multi-bit words. One such IC is the 74HC85 CMOS 4-bit magnitude comparator shown in the below figure. This IC compares two 4-bit words and provides an output on pins 5, 6 and 7 that indicate whether the input words are equal, or if not, whether A or B has the higher numerical value.

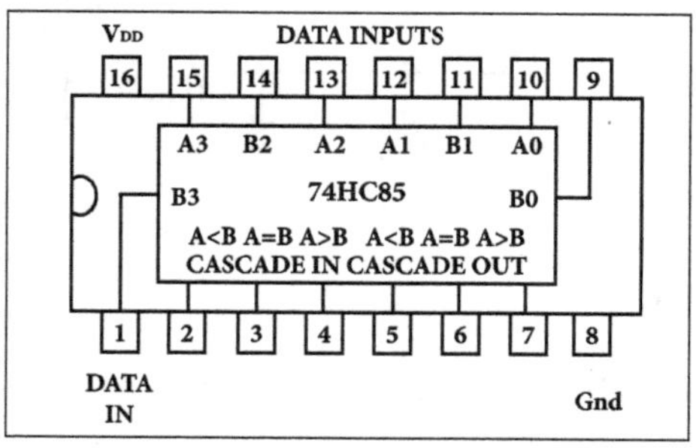

74HC85 4-bit magnitude comparator.

5

Sequential Logic IC's and Memories

5.1 Familiarity with Commonly Available 74xx and CMOS 40xx Series IC's

74 Series Families

The 74LS (Low-power Schottky) family (like the original) uses TTL (Transistor-Transistor Logic) circuitry which is fast but requires more power than later families. The 74 series is often still called the TTL series even though the latest ICs do not use TTL.

The 74HC family has High-speed CMOS circuitry, combining the speed of TTL with the very low power consumption of the 4000 series. They are CMOS ICs with the same pin arrangements as the older 74LS family. Note that 74HC inputs cannot be reliably driven by 74LS outputs because the voltage ranges used for logic 0 are not quite compatible, use 74HCT instead.

The 74HCT family is a special version of 74HC with 74LS TTL-compatible inputs so 74HCT can be safely mixed with 74LS in the same system. In fact 74HCT can be used as low-power direct replacements for the older 74LS ICs in most circuits. The minor disadvantage of 74HCT is a lower immunity to noise, but this is unlikely to be a problem in most situations.

For most new projects the 74HC family is the best choice. The 74LS and 74HCT families require a 5V supply so they are not convenient for battery operation.

Open Collector Outputs

Some 74 series ICs have open collector outputs, this means they can sink current but they cannot source current. They behave like an NPN transistor switch.

The above figure shows how an open collector output can be connected to sink current from a supply which has a higher voltage than the logic IC supply. The maximum load supply is 15V for most open collector ICs.

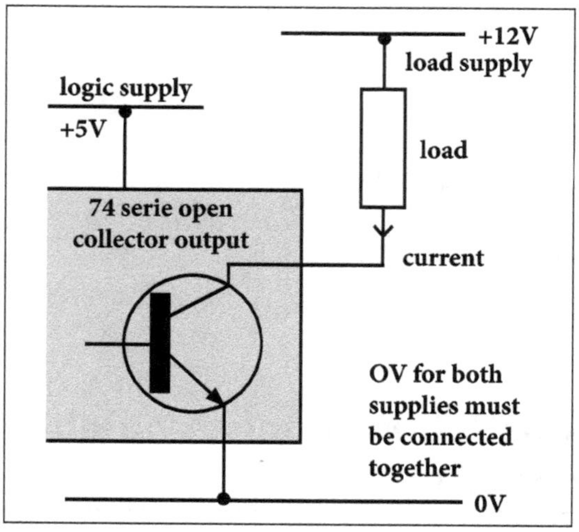

Open Collector Output.

Open collector outputs can be safely connected together to switch on a load when any one of them is low; unlike normal outputs which must be combined using diodes.

74HC and 74HCT Family Characteristics

The CMOS circuitry used in the 74HC and 74HCT series ICs means that they are static sensitive. Touching a pin while charged with static electricity (from our clothes for example) may damage the IC.

In fact most ICs in regular use are quite tolerant and earthing our hands by touching a metal water pipe or window frame before handling them will be adequate. ICs should be left in their protective packaging until we are ready to use them:

- 74HC Supply: 2 to 6V, small fluctuations are tolerated.
- 74HCT Supply: 5V ±0.5V, a regulated supply is best.
- Inputs have very high impedance (resistance), this is good because it means they will not affect the part of the circuit where they are connected. However, it also means that unconnected inputs can easily pick up electrical noise and rapidly change between high and low states in an unpredictable way. This is likely to make the IC behave erratically and it will significantly increase the supply current. To prevent problems all unused inputs must be connected to the supply (either +Vs or 0V), this applies even if that part of the IC is not being used in the circuit.
- Note that 74HC inputs cannot be reliably driven by 74LS outputs because the voltage ranges used for logic 0 are not quite compatible. For reliability use 74HCT if the system includes some 74LS ICs.
- Outputs can sink and source about 4mA, if we wish to maintain the correct output voltage to drive logic inputs, but if there is no need to drive any inputs the maximum current is about 20mA. To switch larger currents we can connect a transistor.

- Fan-out: One output can drive many inputs (50+), except 74LS inputs because these require a higher current and only 10 can be driven.
- Gate propagation time: About 10ns for a signal to travel through a gate.
- Frequency: Up to 25MHz.
- Power consumption (of the IC itself) is very low, a few µW. It is much greater at high frequencies, a few mW at 1MHz for example.

74Ls Family TTL Characteristics

- Supply: 5V ±0.25V, it must be very smooth, a regulated supply is best. In addition to the normal supply smoothing, a 0.1µF capacitor should be connected across the supply near the IC to remove the spikes generated as it switches state, one capacitor is needed for every 4 ICs.
- Inputs float high to logic 1 if unconnected, but do not rely on this in a permanent (soldered) circuit because the inputs may pick up electrical noise. 1mA must be drawn out to hold inputs at logic 0. In a permanent circuit it is wise to connect any unused inputs to +Vs to ensure good immunity to noise.
- Outputs can sink up to 16mA (enough to light an LED), but they can source only about 2mA. To switch larger currents we can connect a transistor.
- Fan-out: one output can drive up to 10 74LS inputs, but many more 74HCT inputs.
- Gate propagation time: about 10ns for a signal to travel through a gate.
- Frequency: up to about 35MHz (under the right conditions).
- Power consumption (of the IC itself) is a few mW.

Quad 2-Input Gates

- 7400 quad 2-input NAND.
- 7403 quad 2-input NAND with open collector outputs.
- 7408 quad 2-input AND.
- 7409 quad 2-input AND with open collector outputs.
- 7432 quad 2-input OR.
- 7486 quad 2-input EX-OR.
- 74132 quad 2-input NAND with Schmitt trigger inputs.

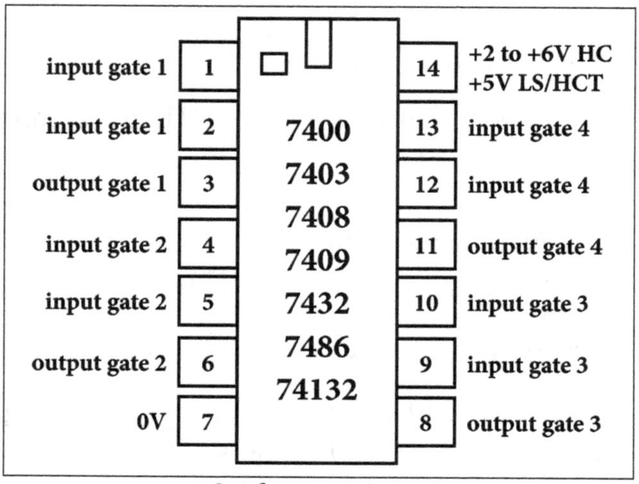

Quad 2-Input Gates.

The 74132 has Schmitt trigger inputs to provide good noise immunity. They are ideal for slowly changing or noisy signals.

7402 Quad 2-input NOR

The 7402 IC is shown separately because it has an unusual gate layout.

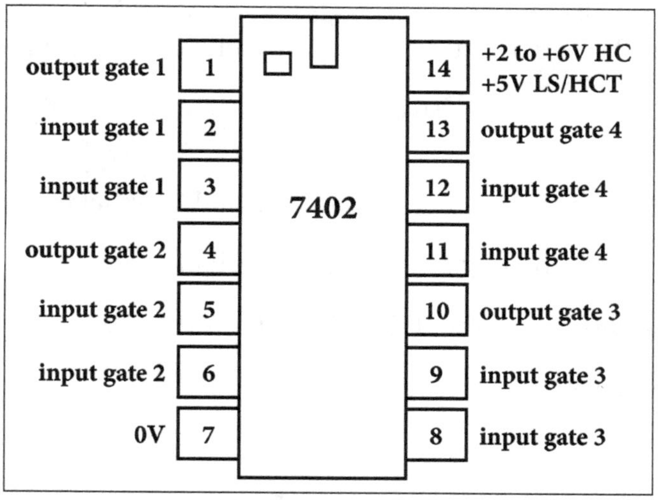

7402 Quad 2-Input NOR Gates.

Triple 3-Input Gates

- 7410 triple 3-input NAND.
- 7411 triple 3-input AND.
- 7412 triple 3-input NAND with open collector outputs.
- 7427 triple 3-input NOR.

Notice how gate 1 is spread across the two sides of the package.

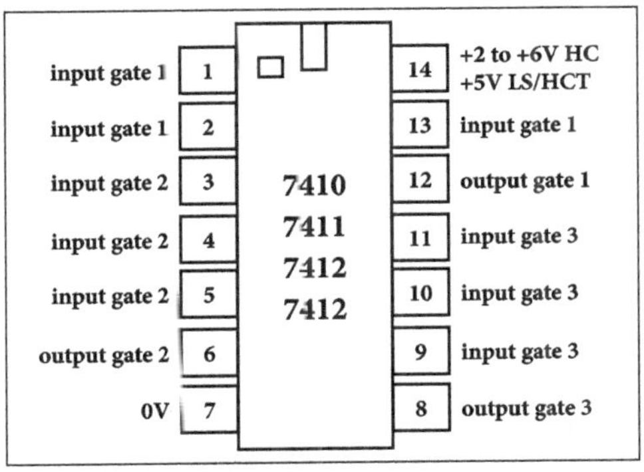

Triple 3-Input Gates.

Dual 4-Input Gates

- 7420 dual 4-input NAND.
- 7421 dual 4-input AND.

NC = No Connection (unused pin).

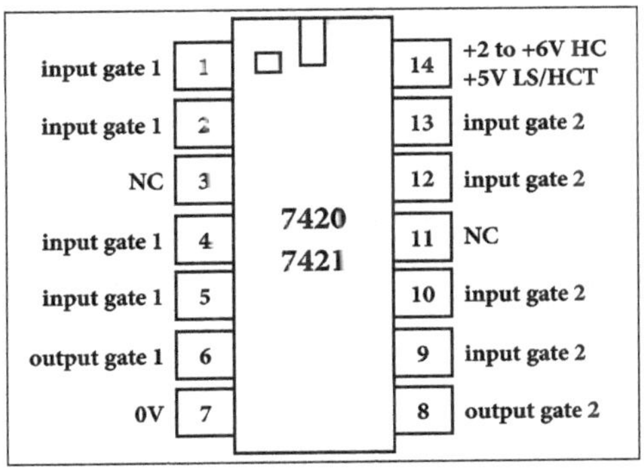

Dual 4-Input Gates.

7430 8-Input Nand Gate

NC = No Connection (unused pin).

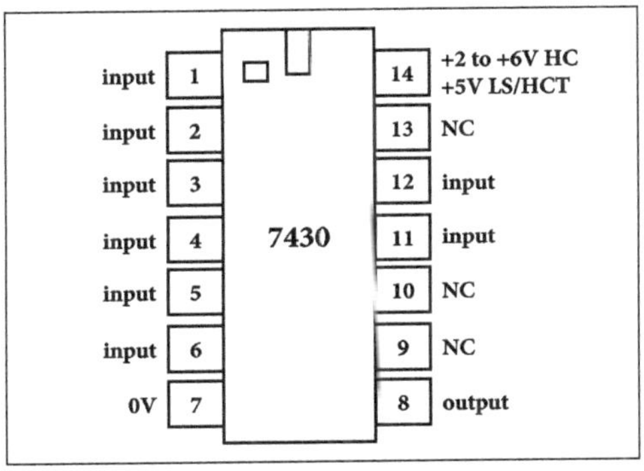

7430 8-input NAND gate.

Hex Not Gates

- 7404 hex NOT.
- 7405 hex NOT with open collector outputs.
- 7414 hex NOT with Schmitt trigger inputs.

The 7414 has Schmitt trigger inputs to provide good noise immunity. They are ideal for slowly changing or noisy signals.

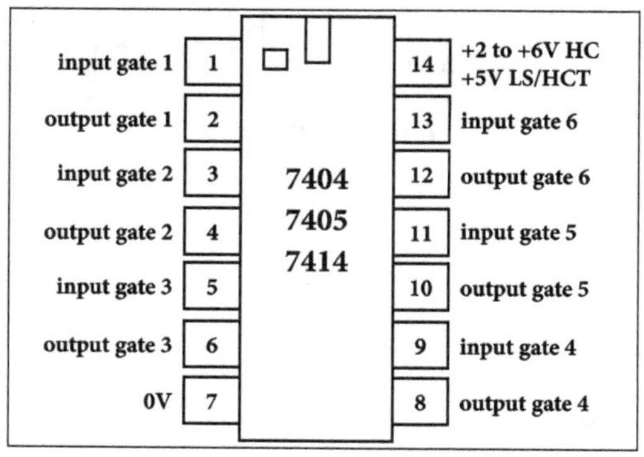

Hex NOT Gates.

7490 Decade (0-9) Ripple Counter and 7493 4-Bit (0-15) Ripple Counter

These are ripple counters so beware that glitches may occur in any logic gate systems connected to their outputs due to the slight delay before the later counter outputs respond to a clock pulse.

NC=No Connection (unused pin).

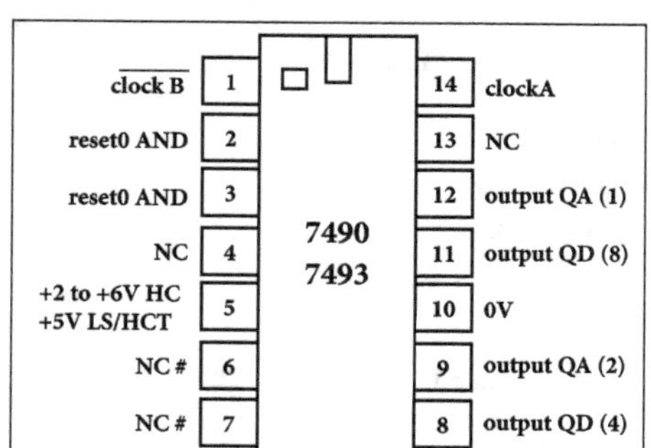

7490 pins 6 and 7 connect to an internal AND gate for resetting to 9.

For normal use connect QA to clock B and connect external clock signal to clock A. The count advances as the clock input becomes low (on the falling-edge), this is indicated by the bar over the clock label. This is the usual clock behaviour of ripple counters and it means a counter output can directly drive the clock input of the next counter in a chain.

The counter is in two sections: clock A-QA and clock B-QB-QC-QD. For normal use connect QA to clock B to link the two sections, and connect the external clock signal to clock A.

For normal operation at least one reset0 input should be low, making both high resets the counter to zero (0000, QA-QD low). Note that the 7490 has a pair of reset 9 inputs on pins 6 and 7, these reset the counter to nine (1001) so at least one of them must be low for counting to occur.

Counting to less than the maximum (9 or 15) can be achieved by connecting the appropriate output(s) to the two reset 0 inputs. If only one reset input is required the two inputs can be connected together. For example: to count 0 to 8 connect QA (1) and QD (8) to the reset inputs.

CMOS 40xx Series IC's

CMOS Logic Families: The first commercially successful CMOS family was 4000-series CMOS. Although 4000-series circuits offered the benefit of low power dissipation, they were fairly slow and were not easy to interface with the most popular logic family of the time, bipolar TTL. Thus, the 4000 series was supplanted in most applications by the more capable CMOS families.

All of the CMOS devices have part numbers of the form 74FAMnn, where "FAM" is an alphabetic family mnemonic and nn is a numeric function designator. Devices in different families with the same value of nn perform the same function. For example, the 74HC30, 74HCT30, 74AC30, 74ACT30, and 74AHC30 are all 8-input NAND gates.

The prefix "74" is simply a number that was used by an early, popular supplier of TTL devices, Texas Instruments. The prefix "54" is used for identical parts that are specified for operation over a wider range of temperature and power-supply voltage, for use in military applications. Such parts are usually, fabricated in the same way as their 74-series counterparts, except that they are tested, screened, and marked differently, a lot of extra paperwork is generated, and a higher price is charged, of course.

$$R\min = 5.0 - 0.0 / IR(\max) = 1562.5$$

$$IR(\text{leak}) = 4 \qquad 5+2 \qquad 20 = 60\ \text{A}$$

$$R\max = 5.0 - 2.4 / IR(\text{leak}) = 43.3$$

HC and HCT

The first two 74-series CMOS families are HC (High-speed CMOS) and HCT (High-speed CMOS, TTL, compatible). Compared with the original 4000 family, HC and HCT both have higher speed and better current sinking and sourcing capability.

The HCT family uses a power supply voltage VCC of 5 V and can be intermixed with TTL devices, which also use a 5-V supply. The HC family is optimized for use in systems that use CMOS logic exclusively and can use any power supply voltage between 2 and 6 V.

A higher voltage is used for higher speed and a lower voltage for lower power dissipation. Lowering the supply voltage is especially effective, since most CMOS power dissipation is proportional to the square of the voltage (CY2f power). Even when used with a 5-V supply, HC devices are not quite compatible with TTL.

VHC and VHCT

Several new CMOS families were introduced in the 1980s and the 1990s. Two of the most recent and probably the most versatile are VHC (Very High-Speed CMOS) and VHCT (Very High-Speed CMOS, TTL compatible). These families are about twice as fast as HC/HCT while maintaining backwards compatibility with their predecessors.

Like HC and HCT, the VHC and VHCT families differ from each other only in the input levels that they recognize; their output characteristics are the same. Also like HC/HCT, VHC/VHCT outputs have symmetric output drive. That is, an output can sink or source equal amounts of current; the output is just as strong in both states. Other logic families, including the FCT and TTL families introduced later, have asymmetric output drive; they can sink much more current in the LOW state than they can source in the HIGH state.

HC, HCT, VHC, and VHCT

Electrical Characteristics Electrical characteristics of the HC, HCT, VHC, and VHCT families are summarized in this subsection. The specifications assume that the devices are used with a nominal 5-V power supply, although (dented) operation is possible with any supply voltage in the range 2-5.5 V (up to 6 V for HC/HCT). Commercial (74-series) parts are intended to be operated at temperatures between 0 C and 70 C, while military (54-series) parts are characterized for operation between 55 C and 125 C.

FCT and FCT-T

In the early 1990s, yet another CMOS family was launched. The key benefit of the FCT (Fast CMOS, 771 compatible) family was its ability to meet or exceed the speed and the output drive capability of the best TTL families while reducing power consumption and maintaining full compatibility with TTL.

5.1.1 All Types of Flip-flops

In electronics, a flip-flop or latch is the circuit which has two stable states and can be used to store the state information. A flip-flop is a bitable multi-vibrator. The circuit can be made to change the state by signals applied to one or more control inputs and that will have one or two outputs. It is an important element of such circuits.

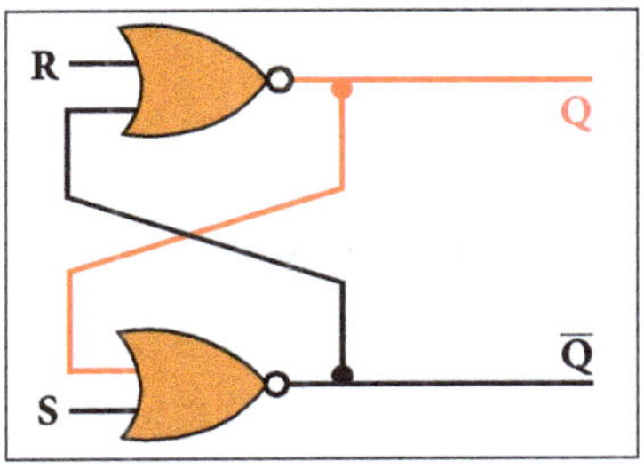

Latch or Flip flop.

The basic unit for storage is flip flop. It maintains its output state either at 1 or 0 till it is directed by an input signal to change its state.

Different Types of Flip-Flop

There are different types of flip flops as follows:

- SR flip-flop.
- D flip-flop.
- JK flip-flop.
- T flip-flop.

SR Flip Flop

The circuit remains in a particular output state indefinitely until something is done to change its output status. If the LOW and HIGH outputs are regarded as 0 and 1 respectively, then the output can either be 0 or 1.

Since either 0 or 1 can be held indefinitely until the circuit is appropriately triggered to go to the other state, the circuit is said to have memory. It is capable of storing one binary digit or one bit of digital information.

Also, in the functioning of the bitable multi-vibrator circuit, when one of the transistors is in saturation, the other will be in cut-off.

This implies that if we had the outputs from the collectors of both the transistors, then the two outputs would be complementary.

In the flip-flops of various types that are available in IC form, provides complementary outputs usually designated as Q and Q'.

The R-S flip-flop is the basic flip-flop. The letters S and R stands for SET and RESET. When the flip-flop is SET, its Q output goes to state 1 and when it is RESET it goes to state 0. The Q output is the complement of the Q output at all times.

Operation of RS flip-flop:

- When the R input is low and the S input is high, the Q output of flip-flop is set.
- When R input is high and S input is low, the Q output of the flip-flop is reset.
- When both the inputs R and S are high, the output is unpredictable.
- When both the inputs R and S are low, the output does not change.

Truth table:

			NAND Gate		
S	R	Q(t + 1)	A	B	Y
0	0	Indeterminate	0	0	1
0	1	Set	0	1	1
1	0	Reset	1	0	1
1	1	NC	1	1	0

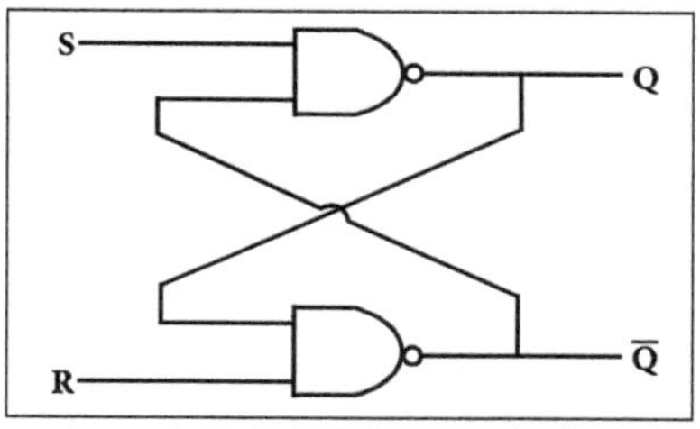

Logic diagram of SR flip-flop.

Latch Flip Flop

The R-S (Reset Set) flip flop is the simplest flip flop of all and easiest to understand. It is basically a device which has two outputs one output being the inverse or complement of the other and two inputs.

A pulse on one of the inputs to take on a particular logical state. The outputs will then remain in this state until a similar pulse is applied to the other input. The two inputs are called the Set and Reset input (sometimes called the preset and clear inputs).

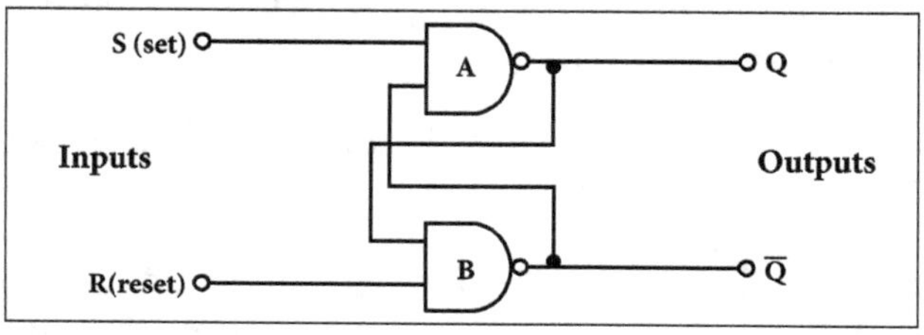

(a) Latch Flip Flop NAND Gate.

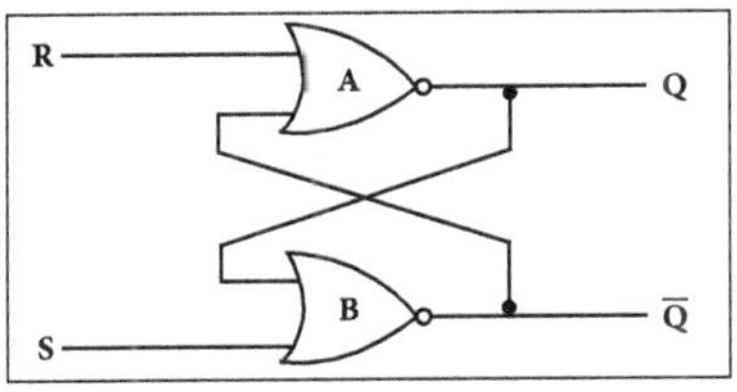

(b) RS Latch Flip Flop NOR Gate.

To describe the circuit of the above figure (a), let us assume that initially both R and S are at the logic 1 state and that output is at the logic 0 state.

Now, if Q = 0 and R = 1, then these are the states of inputs of gate B, therefore the outputs of gate B is at 1 (making it the inverse of Q i.e., 0). The output of gate B is connected to an input of gate A so if S = 1, both inputs of gate A are at the logic 1 state.

This means that the output of gate A must be 0 (as was originally specified). In other words, the 0 state at Q is continuously disabling gate B so that any change in R has no effect. Also the 1 state at (1 is continuously enabling gate A so that any change S will be transmitted through to Q). The above conditions constitute one of the stable states of the device referred to as the Reset state since Q = 0.

Now, if the R-S flip flop in the Reset state, the S input goes to 0. The output of gate i.e., Q will go to 1 and with Q = 1 and R = 1, the output of gates B (0 will go to 0 with Q now 0 gate A is disabled keeping Q at 1). Consequently, when S returns to the 1 state it has no effect on the flip flop whereas a change in R will cause a change in the output of gate B.

The above conditions constitute the other stable state of the device, called the set state since Q = 1. The change of the state of S from 1 to 0 has caused the flip flop to change from the reset state to the set state.

There is another input condition that is, when both the R and S inputs are taken to the logic state 0. When this happens both Q and Q bar will be forced to 1 and will remain so far as long as R and S are kept at 0.

However when both inputs return to 1 there is no way of knowing whether the flip flop will latch in the reset state or the set state. The condition is said to be indeterminate because of this indeterminate state great care must be taken when using R-S flip flop to ensure that both inputs are not instructed simultaneously.

Simple NAND R-S Flip Flop Truth table is described in the following table:

S	R	Q
0	0	indeterminate
0	1	Set(1)
1	0	Reset(0)
1	1	No change

The Truth table for the NAND R-S flip flop is given below:

Initial conditions	Inputs(Pulsed)		Final output	
Q	S	R	Q	
1	0	0	indeterminate	
1	0	1	1	0
1	1	0	0	1
1	1	1	1	0
0	0	0	indeterminate	
0	0	1	1	0
0	1	0	0	1
0	1	1	0	1

When NOR gate are used the R and S inputs are transposed compared with the NAND version. Also the stable state when R and S are both 0.

A change of state is effected by pulsing the appropriate input to the 1 state. The indeterminate state is now when both R and S are simultaneously at logic 1.

NOR Gate R-S Flip Flop Truth table:

S	R	Q
0	0	No change
0	1	Reset(0)
1	0	Set(1)
1	1	Indeterminate

Clocked RS Flip Flop

The RS latch flip flop required the direct input but no clock. It is very useful to add clock to control precisely the time at which the flip flop changes the state of its output.

In the clocked R-S flip flop the appropriate levels applied to their inputs are blocked till the receipt of a pulse from another source called clock. The flip flop changes state only when clock pulse is applied depending upon the inputs.

The basic circuit is shown in the below figure. This circuit is formed by adding two AND gates at inputs to the R-S flip flop. In addition to control inputs Set (S) and Reset (R), there is a clock input (C) also.

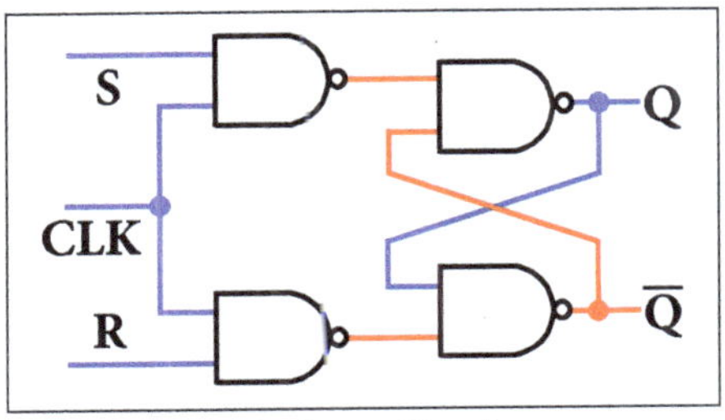

Clocked RS Flip Flop.

The truth tables for the Clocked R-S flip flop:

Initial conditions	Inputs(Pulsed)		Final output	Comment
Q	S	R	Q(t + 1)	No change
0	0	0	0	No change
0	0	1	0	Clear Q
0	1	0	1	Set Q
0	1	1	???	indeterminate
1	0	0	1	No change
1	0	1	0	Clear Q
1	1	0	1	Set Q
1	1	1	???	indeterminate

Excitation table for R-S Flip Flop:

S	R	Q
0	0	No change
0	1	Reset(0)
1	0	Set(1)
1	1	Indeterminate

JK Flip-Flop

The J-K flip-flop behaves in the same fashion as an R-S flip-flop except for one entry in the function table.

In case of R-S flip-flop, its input combination S = R = 1 and the input combination S = R = 0 are prohibited.

In the case of the J-K flip-flop with active HIGH inputs, the output of the flip-flop toggles, (i.e.,) it goes to the other state, for J = K = 1.

The output toggles for J = K = 0 in the case of the flip-flop having an active LOW input. Thus, the J-K flip-flop can overcome the problem of a forbidden input combination of the R-S flip-flop.

The corresponding Karnaugh maps are shown in below figure for the characteristics table.

The characteristic equations is given by,

$$Q_{n+1} = J.\overline{Q_n} + \overline{K}.Q_n$$

$$Q_{n+1} = \overline{J}.\overline{Q_n} + K.Q_n$$

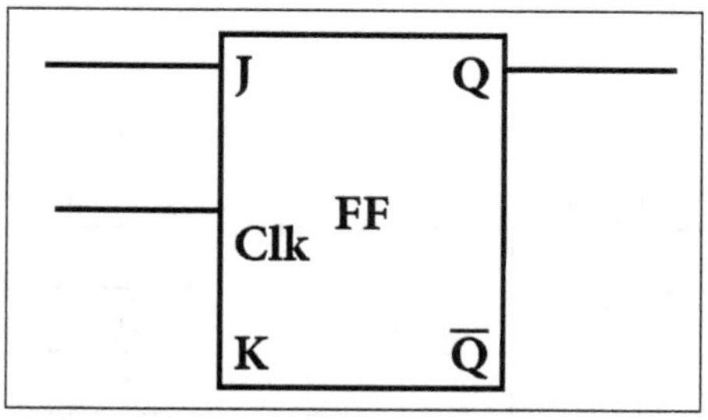

(a) JK flip flop with active high inputs.

Truth table:

Operation mode	J	K	Clk	Q_{n+1}
SET	1	0	1	1
RESET	0	1	1	0
NO CHANGE	0	0	1	Q_n
TOGGLE	1	1	1	Q_n

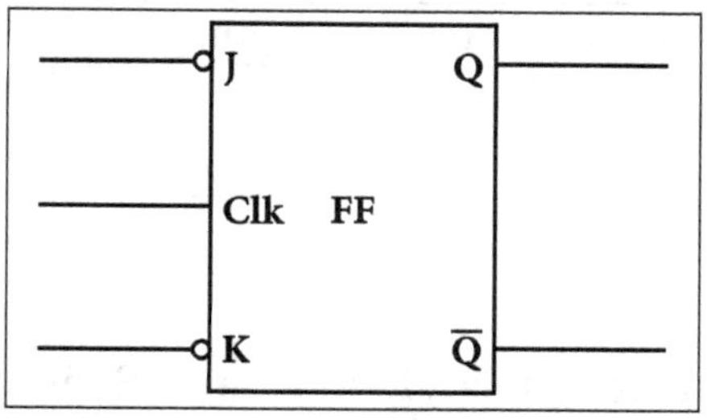

(b) JK flip flop with active low inputs.

Truth table:

Operation mode	J	K	Clk	Q_{n+1}
SET	0	1	1	1
RESET	1	0	1	0
NO CHANGE	1	1	1	Qn
TOGGLE	0	0	1	Q_n

JK Flip-Flop has the following features:

- When J = 1 and K = 1, the output of SIC flip flops toggles (changes to opposite state) on a clock edge.
- When J = 0 and K = 0, the output of JK flip flops does not change on a clock edge.
- When J = 0 and K = 1, the output Q resets to 0 after the clock edge.
- When J = 1 and K =0, the output Q sets to 1 after the clock edge.

Truth table:

Inputs					Outputs		State
J	$\overline{Q}$	CLK	K	Q	Q_{n+1}	$\overline{Q}_{n+1}$	$(n+1)^{th}$
1	0	1 at the +ve edge	1	1	0	1	Toggle $Q_{n+1} = \overline{Q}_n$
1	1	1 at the +ve edge	1	0	1	0	Toggle $Q_{n+1} = \overline{Q}_n$
1	0	–ve edge or 0 to 1	1	1	1	0	Q_n No Change
1	1	–ve edge or 0 to 1	1	1	1	0	Q_n No Change

JK Flip-Flop is a Universal Flip-Flop

When configured in various ways, JK flip flop is capable of operating like most other types of flip-flops.

Operation of JK flip-flop:

- When the K input is low and the J input is high, the Q output of flip-flop is set.
- When the K input is high and the J input is low, the Q output of flip-flop is reset.
- When both the inputs K and J are low, the output does not change.
- When both the inputs K and J are high, it is possible to set or reset the flip-flop (i.e.,) the output toggle on the next positive clock edge.

State Diagram:

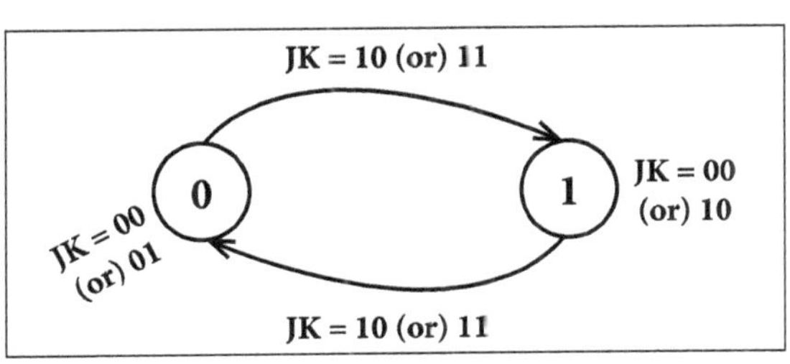

State table:

Q	J	K	Q(t + 1)
0	0	0	0
0	0	1	0
0	1	0	1
0	1	1	1
1	0	0	1
1	0	1	0
1	1	0	1
1	1	1	0

Counter using JK flip-flop:

Q_2	Q_1	Q_0
0	0	0
0	0	1
0	1	1
1	1	1
1	1	0
1	0	0
0	0	0

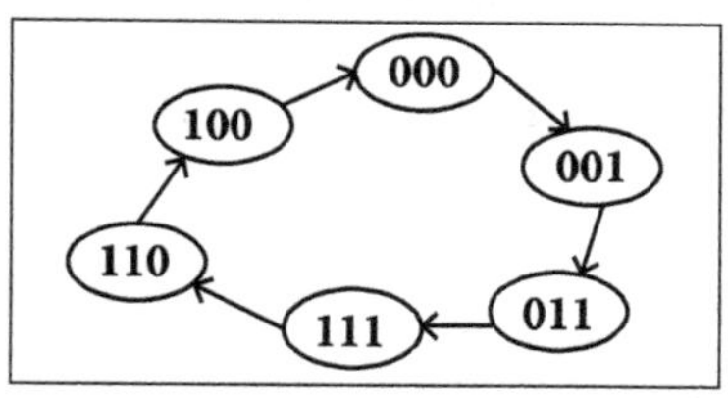

State Diagram.

State table:

PS			NS			JK Input					
Q_2	Q_1	Q_0	Q_{0+1}	Q_{2+1}	Q_{1+1}	J_2	K_2	J_1	K_1	J_0	K_0
0	0	0	1	0	0	0	X	0	X	1	X
0	0	1	1	0	1	0	X	1	X	X	0
0	1	0	0	0	0	0	X	X	1	0	X
0	1	1	1	1	1	1	X	X	0	X	0
1	0	0	0	0	0	X	1	0	X	0	X
1	0	1	0	0	0	X	1	0	X	X	1
1	1	0	0	1	0	X	0	X	1	0	X
1	1	1	0	1	1	X	0	X	0	X	1

K - Map:

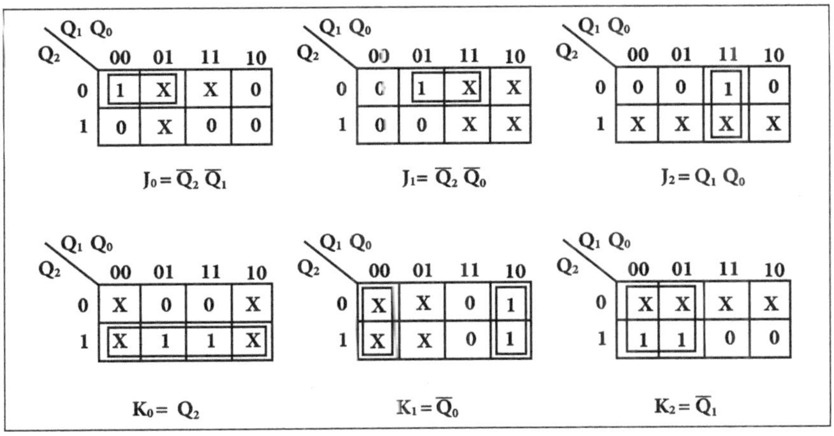

Logic Diagram:

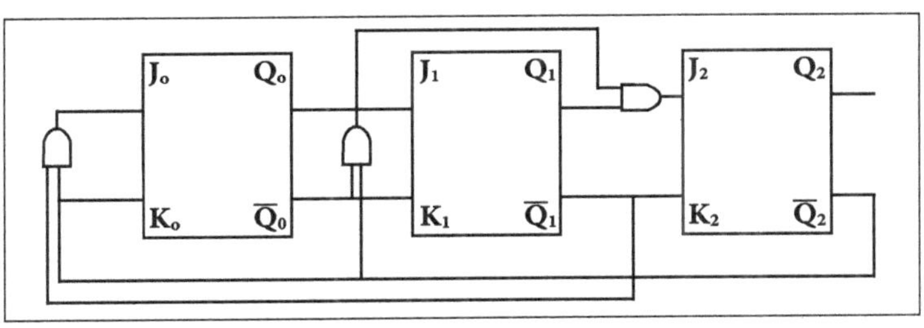

D Flip-Flop

A D flip-flop also called as a delay flip-flop can be used to provide temporary storage of one bit of information.

The below figure shows the symbol of the circuit and the function table of a negative edge-triggered D flip-flop.

When the clock is active, the data bit (0 or 1) present at the D input is transferred to the output.

In the D flip-flop, the data transfer from D input to Q output occurs on the negative-going transition of the clock input. D input can acquire new status.

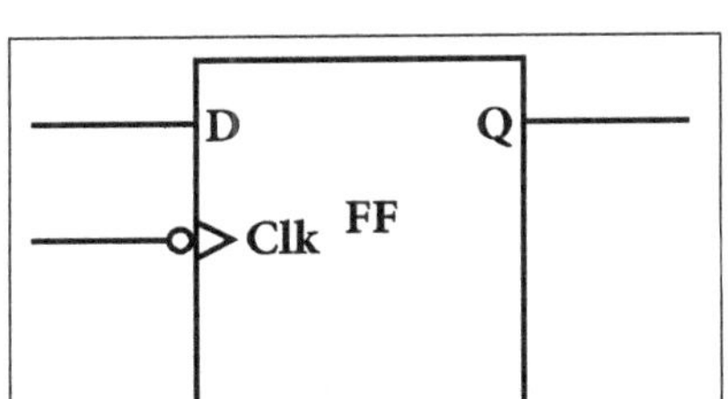

Symbol

Function table:

D	Clk	Q
0	↓	0
1	↓	1

Truth table:

Q_n	D	Q_{n+1}
0	0	0
0	1	1
1	0	0
1	1	1

K-Map:

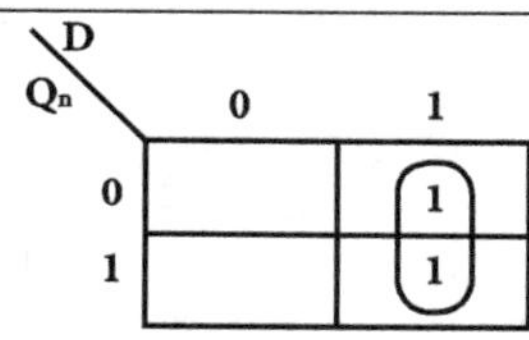

$Q_{n+1} = D$

(d)

Operation of D Flip-Flop

In D flip-flop during the occurrence of clock pulse if D = 1, the output Q is set and if D=0, the output is reset.

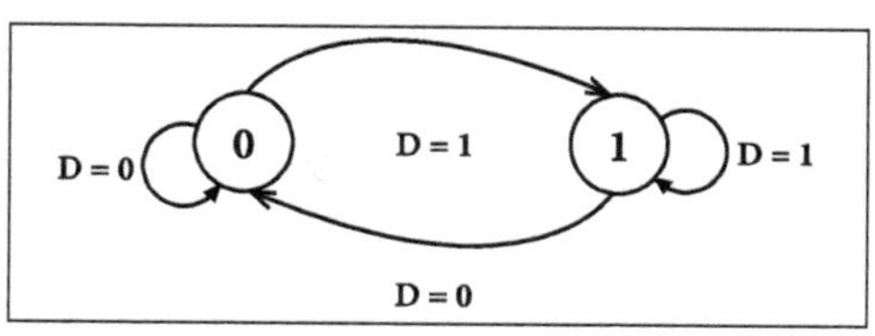

State diagram.

State table:

Q	D	Q(t + 1)
0	0	0
0	1	1
1	0	0
1	1	1

Flip-Flop Excitation Tables for D Flip-Flop

In D flip-flop, its next state is always equal to the D input and it is independent of the present state.

D is 0 if Q_{n+1} has to 0 and if Qn+1 has to be 1 regardless the value of Q_n.

T Flip-Flop

The output of a toggle flip-flop also called a T flip-flop changes state every time it is triggered at its T input known as the toggle input. That is, its output becomes 1, if it was 0 and 0 if it was 1.

The characteristic equations can be written as follows,

$$Q_{n+1} = T \cdot \overline{Q}_n + \overline{T} \cdot Q_n$$

$$Q_{n+1} = \overline{T} \cdot \overline{Q}_n + T \cdot Q_n$$

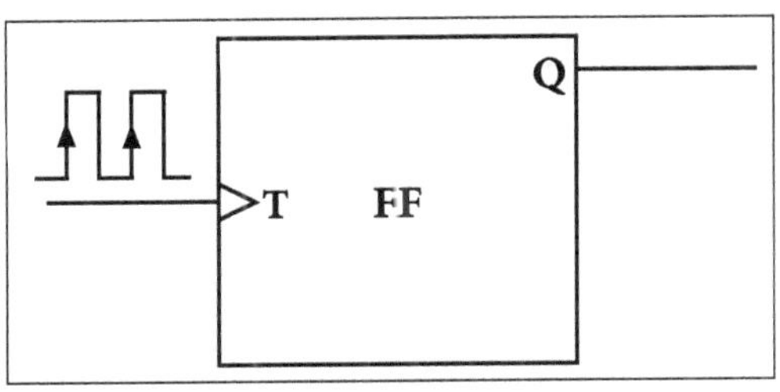

T flip-flop.

T	Q_n	Q_{n+1}
↑	0	1
↑	1	0

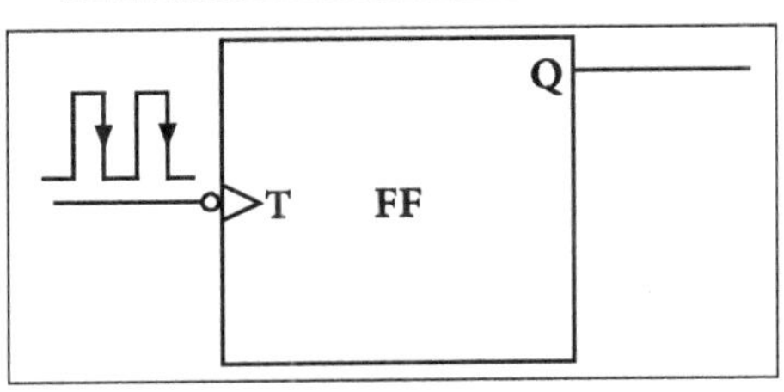

T flip-flop.

T	Q_n	Q_{n+1}
↓	0	1
↓	1	0

Truth table:

Q_n	T	Q_{n+1}
0	0	0
0	1	1
1	0	1
1	1	0

Operation of T Flip-Flop

T flip-flop is also known as Toggle flip-flop.

When the value of T = 0, there is no change in the output.

When the value of T = 1, the output switch to the complement state (i.e.,) the output toggles.

Flip-Flop Excitation Tables for T Flip-Flop

When input T = 1, the state of the flip-flop is complemented.

When T = 0, the state of the flip-flop remains unchanged. Therefore, for 00 and 11 transitions T must be 0 and for 01 and 10 transitions T must be 1.

Master-Slave Jk Flip-Flop

A master-slave flip-flop has two flip-flops where one circuit serves as master and the other as slave.

The Master-Slave Flip-Flop is basically two gated SR flip-flops that is connected together in a series configuration with slave having an inverted clock pulse.

The outputs from the Q and Q from the "Slave" flip-flop are fed back to inputs of "Master" with outputs of "Master" flip-flop being connected to two inputs of the "Slave" flip -flop. This feedback configuration from slave's output to the master's input gives characteristic toggle of J-K flip-flop as shown below:

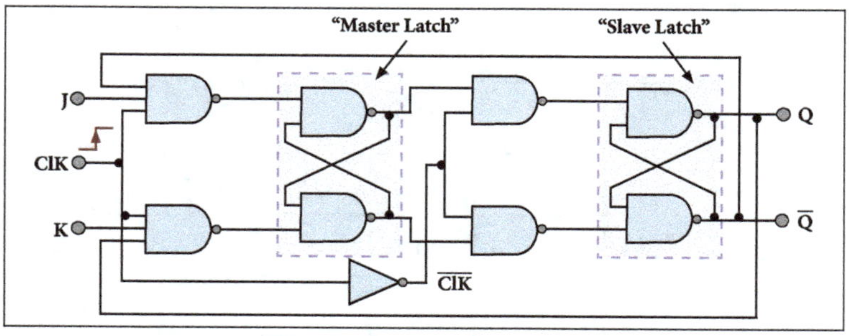

Master-Slave JK Flip-Flop.

The input signals J and K are connected to the gated "master" SR flip flop that "locks" the input condition while clock (Clk) input is "HIGH" at the logic level "1".

As the clock input of the "slave" flip flop is the inverse of "master" clock input, the "slave" SR flip flop does not toggle.

The outputs from "master" flip flop are only "seen" by gated "slave" flip flop when clock input goes "LOW" to logic level "0".

When clock is "LOW", the outputs from "master" flip flop are latched and additional changes to its inputs are being ignored.

The gated "slave" flip flop now responds to the state of its inputs passed over by the "master" section.

Then, on "Low-to-High" transition of clock pulse, inputs of the "master" flip flop are fed to the gated inputs of "slave" flip flop and on "High-to-Low" transition, the same inputs are reflected on output of the "slave" making this type of flip flop edge or pulse-triggered.

Then, circuit accepts input data when clock signal is "HIGH" and passes data to the output on the falling-edge of clock signal.

In other words, Master-Slave JK Flip flop is a "Synchronous" device as it only passes the data with the timing of clock signal.

Application of Master Slave FF:

- It can be used as frequency dividers.
- Master slave FF's can be used as Shift registers.
- Parallel data storage.
- It is used in counters.

5.1.2 Synchronous Counters

Synchronous counters are so called because the clock input of all the individual flip-flops within the counter are all clocked together at the same time by the same clock signal.

The counters which use a clock signal to change their transition are called "Synchronous counters". This means the synchronous counters depends on their clock input to change state values. In synchronous counters, all flip flops are connected to the same clock signal and all flip flops will trigger at the same time.

Synchronous counters are also known as 'Simultaneous counters'. There is no propagation delay and no ripple effect in synchronous counters.

Design of Synchronous Counters

Procedure to design a synchronous counter:

- Step 1: Obtain the truth table of the logic sequence for intended counter to be designed. Alternatively obtain the state diagram of the counter.
- Step 2: Determine number and type of flip-flop to be used.
- Step 3: From the excitation table, determine its next state logic.
- Step 4: From the output state, use Karnaugh map for simplification to derive the circuit output functions and flip-flop output functions.
- Step 5: Draw the logic circuit diagram.
- Step 6: Simulate the circuit using software.
- Step 7: Build the circuit.

State Diagram

A state represents the status of the flip flops in a sequential circuit. If there are n flip flops then the circuit can have 2n possible states.

State diagram is the graphical representation of the information available in a state table. In the state diagram, a state is represented by a circle and the transitions between states are indicated by directing lines connecting the circles.

In addition to graphical symbols, tables or equations, flip-flops can also be represented graphically by a state diagram.

In this diagram, a state is represented by a circle and the transition between states is indicated by the directed lines (or arcs) connecting the circles. An example of a state diagram is shown in the below figure:

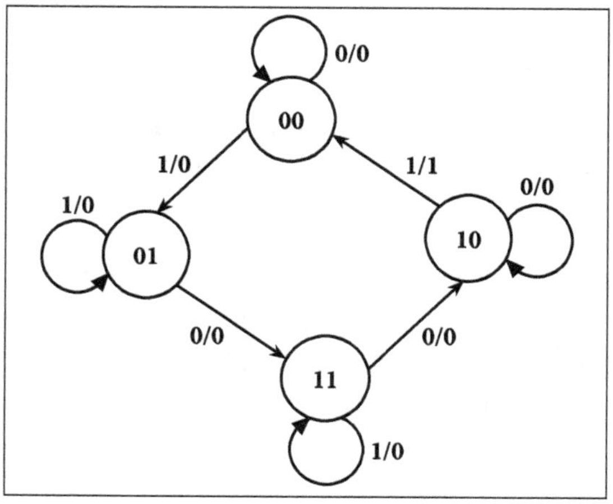

State diagram.

State Table

State table representation of a sequential circuit consists of three sections labelled present state, next state and output.

The present state designates the state of flip-flops before occurrence of a clock pulse. The next state shows the states of flip-flops after the clock pulse and the output section lists the value of the output variables during the present state.

A state table gives the time sequence of inputs, outputs and flip-flop states. A table consists of four sections labelled as present state, next state, input and output.

State Minimization

A partition consists of one or more blocks, where each block comprises a subset of states that may be equivalent, but states in a given block are definitely not equivalent to the states in other blocks.

State Minimization through Partitioning:

- Form an initial partition (P_1) which includes all states.
- Form a second partition (P_2) by separating the states into two blocks based on their output values.
- Form a third partition (P_3) by separating the states into blocks corresponding to next state values.
- Continue partitioning until two successive partitions are the same (i.e. $P_{N-1} = P_N$).
- All states in any one block are equivalent.
- Equivalent states can be combined into a single state.

State Assignment

The cost of the combinational circuit of a sequential circuit can be reduced by using the known simplification methods for combinational circuits.

State assignment procedures are concerned with methods for assigning binary values to states in such a way as to reduce the cost of combinational circuit that drives the flip - flops.

Techniques used for state assignment:

- Shared row state assignment.
- One hot state assignment.

Excitation Table and Maps

A synchronous decade counter counts from zero to nine and repeat the sequence.

Since there are ten states, four JK flip-flops are required. The truth tables of present and next state for the decade counter are shown in the below figure:

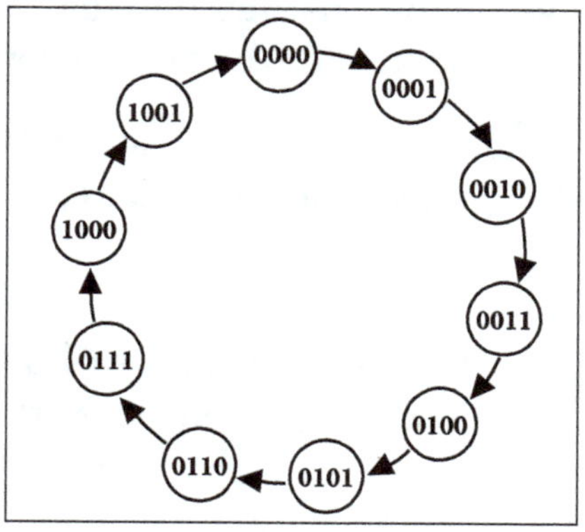

State diagram of synchronous decade counter.

Truth table and state table of synchronous decade counter:

Present State				Next State				Output							
Q_3	Q_2	Q_1	Q_0	Q_3	Q_2	Q_1	Q_0	J_3	K_3	J_2	K_2	J_1	K_1	J_0	K_0
0	0	0	0	0	0	0	1	0	X	0	X	0	X	1	X
0	0	0	1	0	0	1	0	0	X	0	X	1	X	X	1
0	0	1	0	0	0	1	1	0	X	0	X	X	0	1	X
0	0	1	1	0	1	0	0	0	X	1	X	X	1	X	1
0	1	0	0	0	1	0	1	0	X	X	0	0	X	1	X
0	1	0	1	0	1	1	0	0	X	X	0	1	X	X	1
0	1	1	0	0	1	1	1	0	X	X	0	X	0	1	X
0	1	1	1	1	0	0	0	1	X	X	1	X	1	X	1
1	0	0	0	1	0	0	1	X	0	0	X	0	X	1	X
1	0	0	1	0	0	0	0	X	1	0	X	0	X	X	1

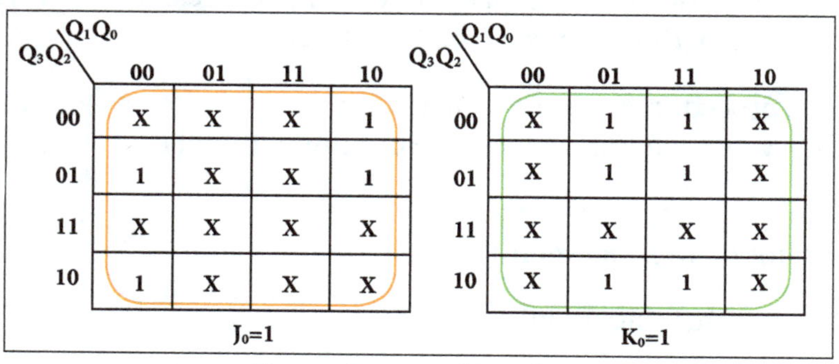

Karnaugh maps of J_0 and K_0.

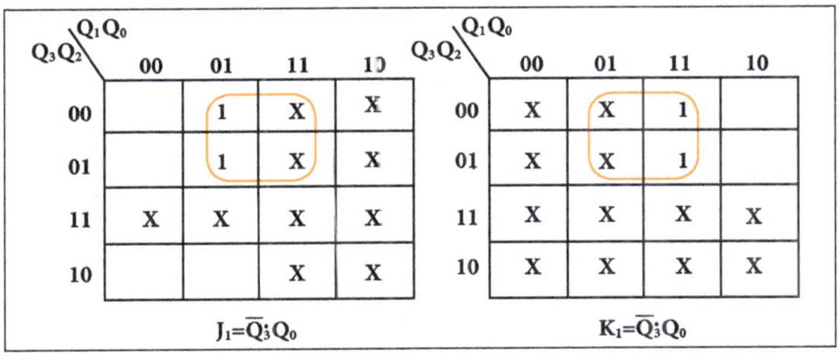

Karnaugh maps of J_1 and K_1.

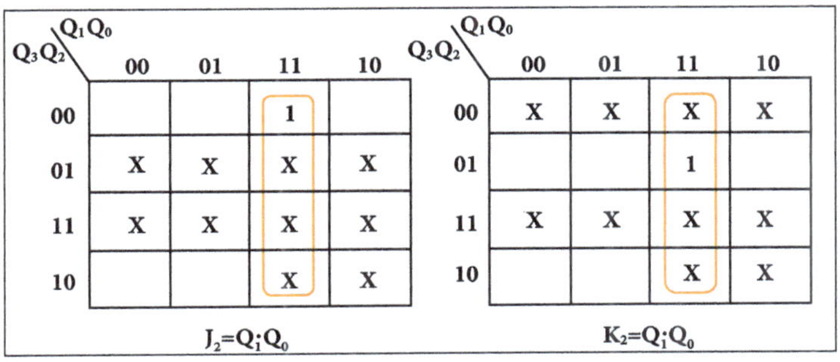

Karnaugh maps of J_2 and K_2.

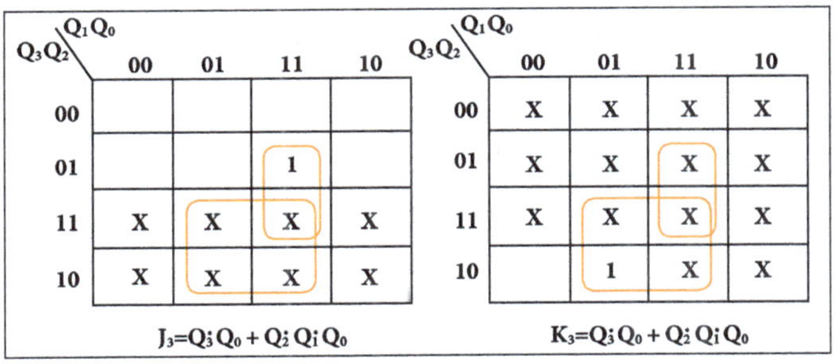

Karnaugh maps of J_3 and K_3.

Circuit Implementation

Based on the results obtained from the Karnaugh maps, the circuit design of synchronous decade counter is shown in the below figure.

The steps to design a Synchronous Counter using JK flip flops are:

The general sequential circuit in terms of its basic parts and its input and outputs.

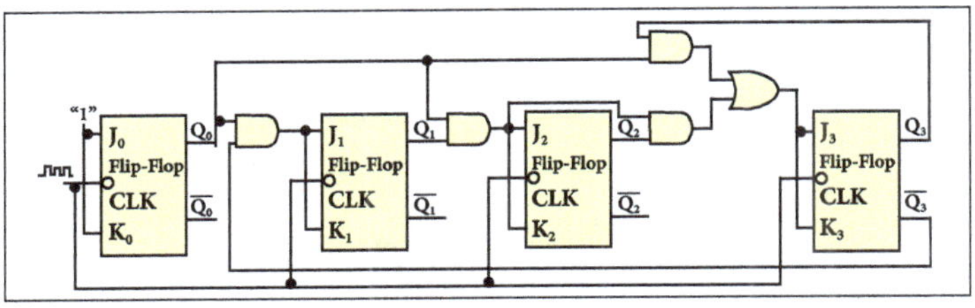

Design of Synchronous decade counter using JK flip-flop.

Let us design a 2 bit up/down counter with an input D which determines the up/down function. Thus, when D = 0, the count sequence is 00, 01, 10, 11, 00... When D = 1, the count sequence is 00, 11, 10, 01, 00...

Solution:

Step 1: Draw the state diagram for the given sequence.

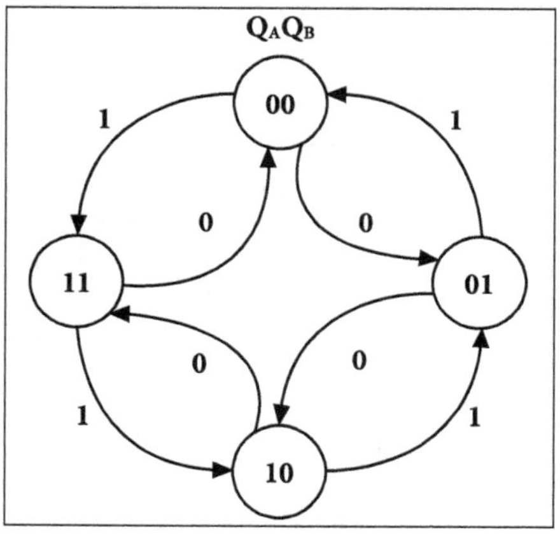

State diagram.

Step 2: Develop a next-state table for the specific counter sequence by using the state diagram as a reference and fill up the present state and next state columns. For this interactive table, we can modify the next state.

Present State			Next State		JK flip flop inputs			
D	QA	QB	QA	QB	JA	KA	JB	KB
0	0	0	0	1	0	X	1	X
0	0	1	1	0	1	X	X	1
0	1	0	1	1	X	0	1	X
0	1	1	0	0	X	1	X	1
1	0	0	1	1	1	X	1	X
1	0	1	0	0	0	X	X	1
1	1	0	0	1	X	1	1	X
1	1	1	1	0	X	0	X	1

JK Flip Flop Truth table:

J	K	Q
0	0	Q_o
0	1	0
1	0	1
1	1	Toggle

JK Flip Flop Transition table:

Q_N	Q_{N+1}	J	K
0	C	0	X
0	1	1	X
1	0	X	1
1	1	X	0

Step 3: Use K-map to derive the logic equations.

$JA = D\ QB + D\ QB$:

QAQB\D	0	1
00	0	1
01	1	0
11	X	X
10	X	X

$KA = D\ QB + D\ QB$:

QAQB\D	0	1
00	X	X
01	X	X
11	1	0
10	0	1

$JB = 1$:

QAQB\D	0	1
00	1	1
01	X	X
11	X	X
10	1	1

$KB = 1$:

QAQB\D	0	1
00	X	X
01	1	1
11	1	1
10	X	X

Step 4: Use the Boolean expressions to implement the counter.

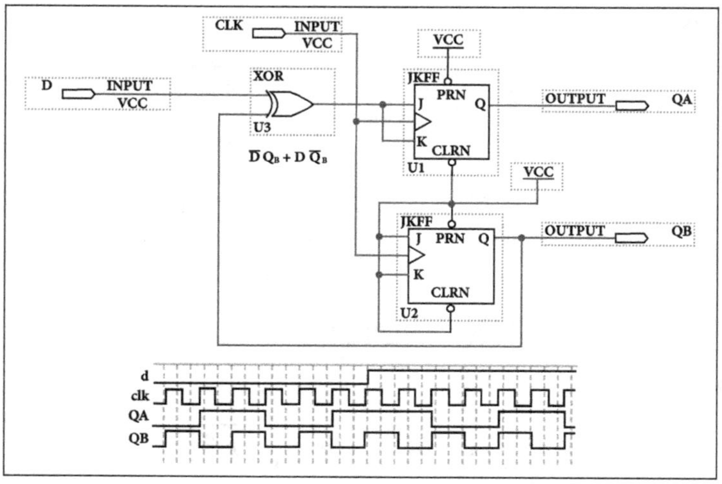

5.1.3 Decade Counters

A synchronous decade counters counts from zero to nine and repeat the sequence. Since there are ten states, four JK flip-flops are required. The truth tables of present and next state for the decade counter are shown in the below figure:

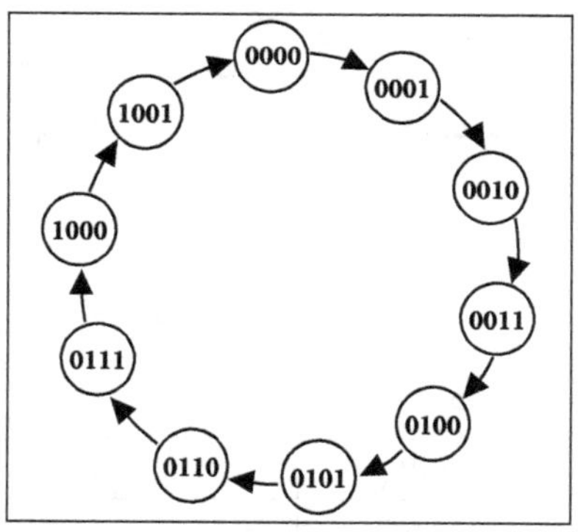

State Diagram of Synchronous Decade Counter.

Truth table and state table of synchronous decade counter:

Present state				Next state				Output							
Q_3	Q_2	Q_1	Q_0	Q_3	Q_2	Q_1	Q_0	J_3	K_3	J_2	K_2	J_1	K_1	J_0	K_0
0	0	0	0	0	0	0	1	0	X	0	X	0	X	1	X
0	0	0	1	0	0	1	0	0	X	0	X	1	X	X	1
0	0	1	0	0	0	1	1	0	X	0	X	X	0	1	X
0	0	1	1	0	1	0	0	0	X	1	X	X	1	X	1
0	1	0	0	0	1	0	1	0	X	X	0	0	X	1	X
0	1	0	1	0	1	1	0	0	X	X	0	1	X	X	1
0	1	1	0	0	1	1	1	0	X	X	0	X	0	1	X

0	1	1	1	1	0	0	0	1	X	X	1	X	1	X	1
1	0	0	0	1	0	0	1	X	0	0	X	0	X	1	X
1	0	0	1	0	0	0	0	X	1	0	X	0	X	X	1

Karnaugh Maps of J_0 And K_0

Q_3Q_2 \ Q_1Q_0	00	01	11	10
00	1	X	X	1
01	1	X	X	1
11	X	X	X	X
10	1	X	X	X

$J_0 = 1$

Q_3Q_2 \ Q_1Q_0	00	01	11	10
00	X	1	1	X
01	X	1	1	X
11	X	X	X	X
10	X	1	X	X

$K_0 = 1$

Karnaugh maps of J_1 and K_1

Q_3Q_2 \ Q_1Q_0	00	01	11	10
00		1	X	X
01		1	X	X
11	X	X	X	X
10			X	X

$J_1 = \overline{Q_3}Q_0$

Q_3Q_2 \ Q_1Q_0	00	01	11	10
00	X	X	1	
01	X	X	1	
11	X	X	X	X
10	X	X	X	X

$K_1 = \overline{Q_3}Q_0$

Karnaugh maps of J_2 and K_2

Q_3Q_2 \ Q_1Q_0	00	01	11	10
00			1	
01	X	X	X	X
11	X	X	X	X
10			X	X

$J_2 = Q_1Q_0$

Q_3Q_2 \ Q_1Q_0	00	01	11	10
00	X	X	1	X
01			1	
11	X	X	X	X
10			X	X

$K_2 = Q_1Q_0$

Karnaugh maps of J_3 and K_3

Q_3Q_2 \ Q_1Q_0	00	01	11	10
00				
01			1	
11	X	X	X	X
10	X	X	X	X

$J_3 = Q_3Q_0 + Q_2Q_1Q_0$

Q_3Q_2 \ Q_1Q_0	00	01	11	10
00	X	X	X	X
01	X	X	X	X
11	X	X	X	X
10		1	X	X

$K_3 = Q_3Q_0 + Q_2Q_1Q_0$

Circuit Implementation

Based on the results obtained from the Karnaugh maps, the circuit design of synchronous decade counter is shown in the figure.

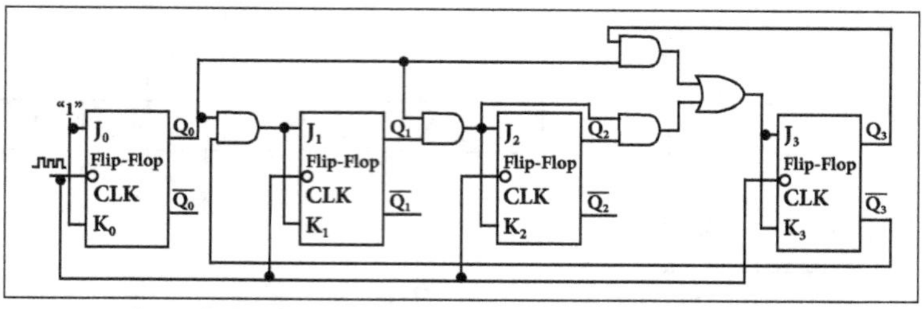

Design of Synchronous decade counter using JK flip-flop.

5.1.4 Shift Registers

The register, which is capable of shifting its binary information in one or both the directions from a state to a state within the register or into or out of the register upon the application of clock pulses, is known as a shift register.

This register is capable of shifting the data, one bit at a time. It is used for converting between the serial and parallel interfaces. It is useful for as many as circuits to work in groups of bits in parallel, but comparatively the serial interfaces are simpler to construct.

The shift registers are used as simple delay circuits and it can also be used as pulse extenders. The serial shift register consists of a chain of flip-flops which is connected in cascade with the output of one flip-flop being connected to the input of its neighbor.

The basic operation of the shift register is synchronous. In shift registers, every flip-flop is connected to a common clock.

The below figure gives the symbolical representation of the different types of data movement in shift register operations.

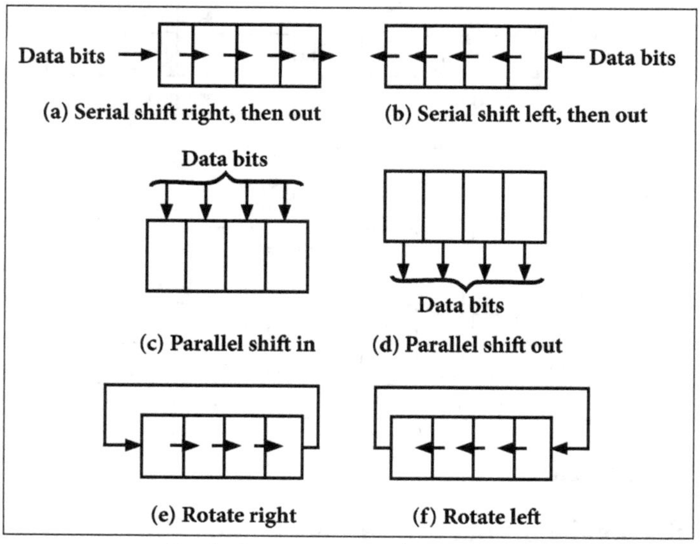

Basic data movement in registers.

The D flip-flops form the simplest type of shift-registers. This basic data movements is possible only within a four-bit shift register is shown in the below figure:

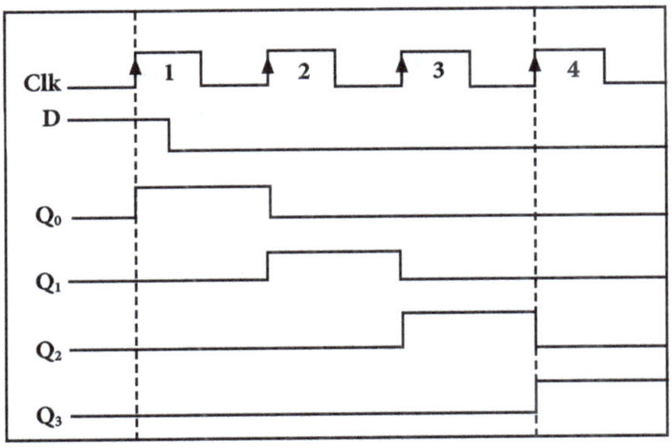

Timing diagram of shift register.

The different types of shift registers are:

- Serial In Serial Out (SISO).
- Serial In Parallel Out (SIPO).
- Parallel In Serial Out (PISO).
- Parallel In Parallel Out (PIPO).

Serial in Serial Out (SISO)

Serial In Serial Out (SISO) shift registers are a kind of shift registers where both data loading as well as data retrieval to/from the shift register occurs in serial-mode. The below figure shows a n-bit synchronous SISO shift register sensitive to positive edge of the clock pulse. Here the data word which is to be stored is fed bit-by-bit at the input of the first flip-flop.

Further, it is seen that the inputs of all other flip-flops (except the first flip-flop FF1) are driven by the outputs of the preceding ones say for example, the input of FF2 is driven by the output of FF1. At last the data stored within the register is obtained at the output pin of the nth flip-flop in serial-fashion.

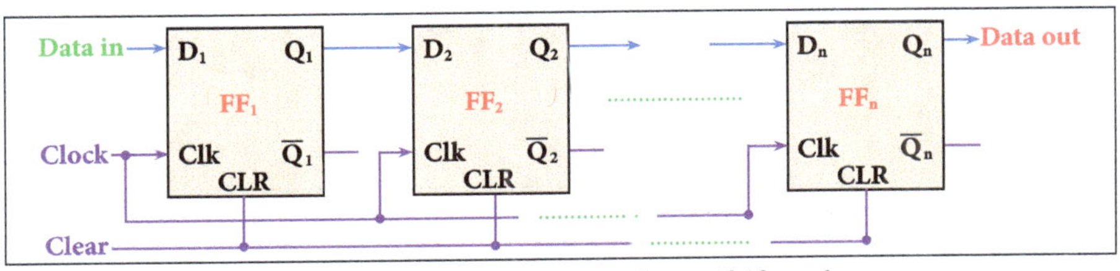

n-bit Right-shift Serial In Serial Out Shift register.

Initially all the flip-flops in the register are cleared by applying high on their clear pins. Next the input data word is fed serially to FF1.

This causes the bit appearing at the D1 pin (B1) to be stored into FF1 as soon as the first leading edge of the clock appears. Further at the second clock tick, B1 gets stored into FF2 while a new bit enters into $FF_1(B_2)$.

This kind of shift in data bits continues for every rising edge of the clock pulse. This indicates that for every single clock pulse the data within the register moves towards right by a single bit. Thus the design shown in the above figure is regarded as a right-shift SISO shift register.

Following the data transmission as explained, one can note that the first bit of an input word appears at the output of nth flip-flop for the nth clock tick. On applying further clock cycles, one gets the next successive bits of the input data word as the serial output (Table 1). The waveforms pertaining to the same are shown by the below figure:

Clock Cycle	Data in	Q_1	Q_2	...	Q_n = Data out
1	B_1	B_1	0	...	0
2	B_2	B_2	B_1	...	0
3	B_3	B_3	B_2	...	0
4	B_4	B_4	B_3	...	0
5	B_5	B_5	B_4	...	0
6	B_6	B_6	B_5	...	0
.	.	.	.	...	.
n	B_n	B_n	B_{n-1}	...	B_1
$n+1$	B_{n+1}	B_{n+1}	B_n	...	B_2
.	.	.	.	...	.

Serial Output Bits of SISO (Right-Shift) Shift Register

Data movement in Right- Shift SISO shift.

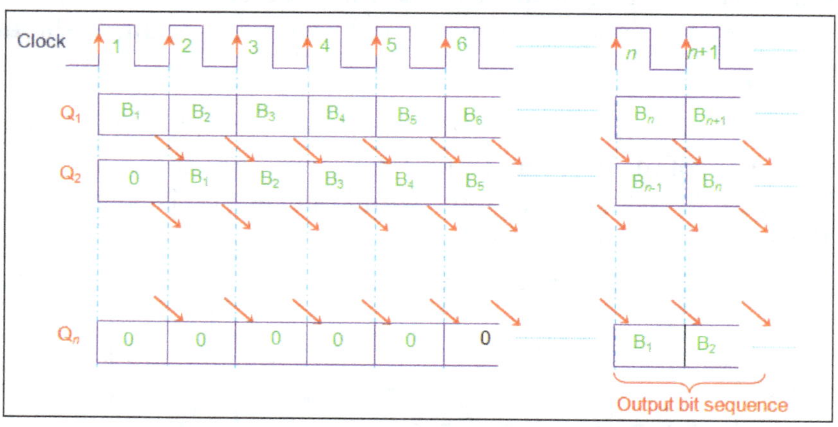

Output waveform of n-bit Right-Shift SISO shift register.

Similar to the right-shift SISO shift-register shown, there can exist a left-shift SISO shift-register also (as shown in the below figure). However the working principle remains the same except the fact that the data movement will be from right to left.

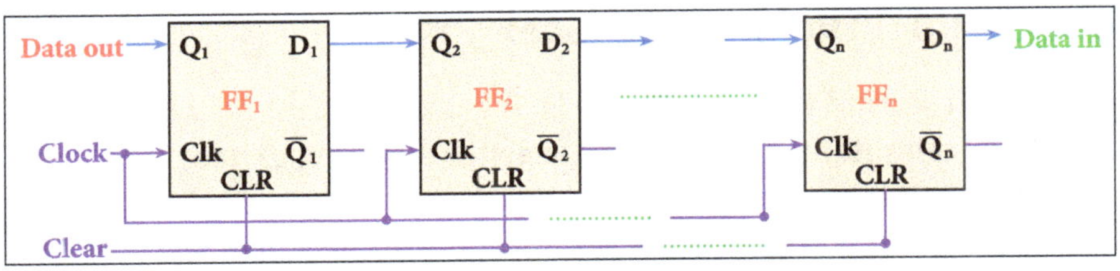

n-bit Left-shift Serial-in Serial out shift register.

Serial in Parallel Out Shift Register

In this case, the data bits are entered into the register in the same manner, i.e. serially. But the output is taken in parallel. Once the data are stored, each bit appears on its respective output line and all bits are available simultaneously, instead of on a bit-by-bit basis as with the serial output as shown in the below figure:

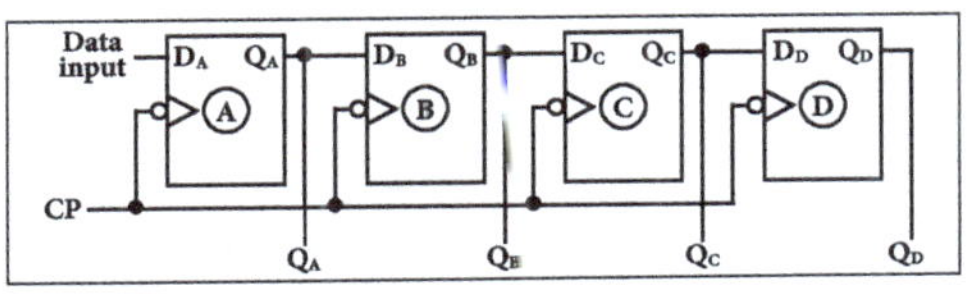

A serial in parallel out shift register.

Parallel In Serial Out Shift Register

In this type, the bits are entered in parallel i.e simultaneously into their respective stages on parallel lines.

The below figure illustrates a four-bit parallel in serial out register. There are four input lines XA, XB, XC, XD for entering data in parallel into the register. $SHIFT/\overline{LOAD}$ is the control input which allows shift or loading data operation of the register.

When $SHIFT/\overline{LOAD}$ is low gates G1, G2, G3 are enabled, allowing each input data bit to be applied to D input of its respective flip-flop. When a clock pulse is applied, the flip-flops with D = 1 will SET and those with D = 0 will RESET. Thus, all four bits are stored simultaneously.

When $SHIFT/\overline{LOAD}$ is high, gates G1, G2, G3 are disabled and gates G4, G5, G6 are enabled. This allows the data bits to shift left from one stage to the next. The OR gates at the D• inputs of the flip-flops allow either the parallel data entry operation or shift operation, depending on which AND gates are enabled by the level on the $SHIFT/\overline{LOAD}$ input.

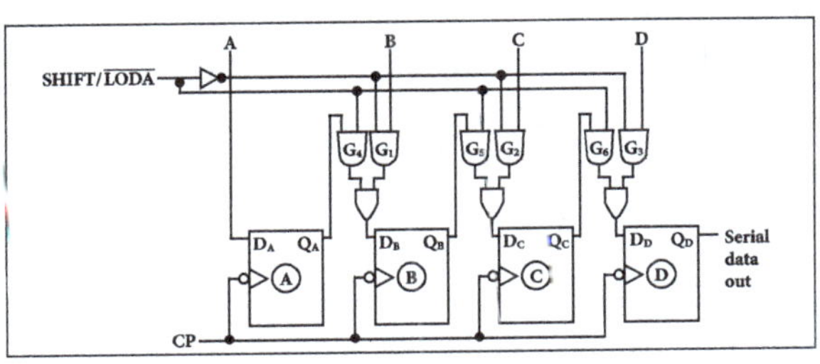

Parallel in serial out shift register.

Parallel in Parallel Out Shift Register

From the third and second types of registers, it is cleared that how to enter the data in parallel i.e. all bits simultaneously into the register and how to take data out in parallel from the register. In 'parallel in parallel out register', there is simultaneous entry of all data bits and the bits appear on parallel outputs simultaneously. The below figure shows this type of register.

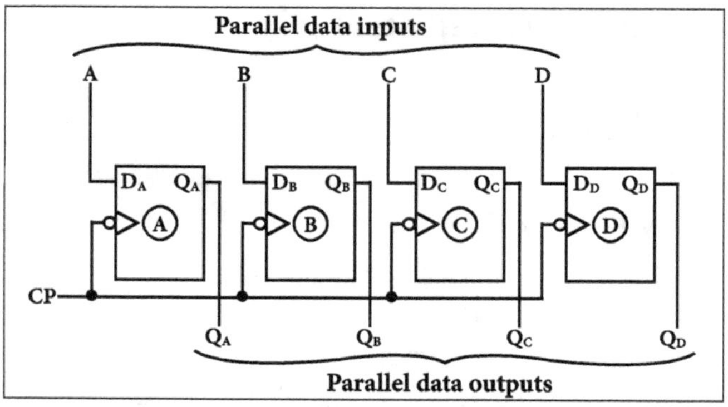

Parallel in parallel out shift register.

Bidirectional Shift Register

The bidirectional shift register allows the shifting of data either to the left or to the right side. This can be implemented by using the logic gate circuitry which enables the transfer of data from one stage to the next stage where either to the right or to the left, depending on the level of a control line. The below figure shows a four-bit bidirectional register.

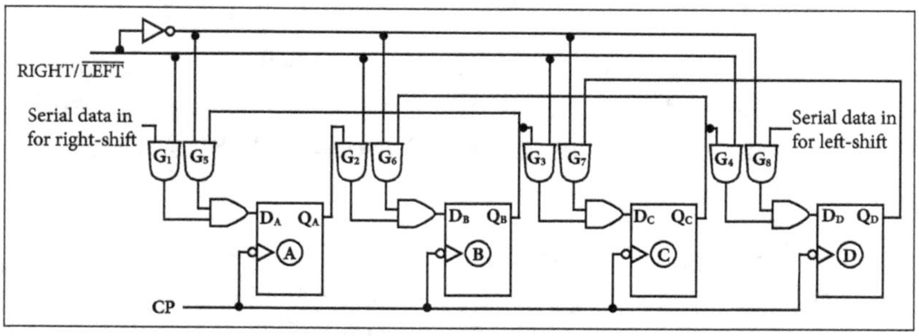

(a) 4-bit bi-directional shift register.

The RIGHT or LEFT is the control input signal which allows the data shifting either towards right or towards left. If high on this line, it enables the shifting of data towards the right and a low enables it towards the left. When the RIGHT or LEFT signal is high, the gates G1, G2, G3, G4 are enabled. The state of the Q output of each flip-flop is passed through the D input of the following flip-flop.

During the arrival of clock pulse, the data are shifted from one place to the right. If the RIGHT or LEFT signal is low, the gates G_5, G_s, G_7, G_5 are enabled. The Q output of each flip-flop is passed through the D input of the preceding flip-flop. During the arrival of clock pulse, the data are shifted from one place to the left. The Bidirectional Shift Register with the Parallel Load we can see that shift register can be used for converting the serial data into parallel data and vice-versa.

While the parallel load capability is added to the shift register, the data entered in parallel can be taken out in serial way by shifting the data stored in the register. This register is known as the bidirectional shift register with parallel load. The figure below shows bidirectional shift register with parallel load.

As shown in the below figure (b), the D input of each flip-flop has three sources:

- Output of left adjacent flip-flop.
- Output of right adjacent flip-flop.
- Parallel input.

From these three sources one of the sources is selected at a time and it is done with the help of the decoder. The decoder selects the lines (SL1 and SL0) by selecting the one source out of three as shown in the table below:

SL_1	SL_0	Selected source
0	0	Parallel input
0	1	Output of right adjacent FF
1	0	Output of left adjacent FF
1	1	-

If the select lines are 00 (i.e., $SL_1 = 0$ and $SL_0 = 0$), the data from the parallel inputs is loaded into the 4-bit register. If the select lines are 01 (i.e., SL1 = 0 and SL0 = 1), the data within the register is shifted 1-bit left. If the select lines are 10 (i.e., $SL_1 = 1$ and $SL_0 = 0$), the data within the register is shifted 1-bit right.

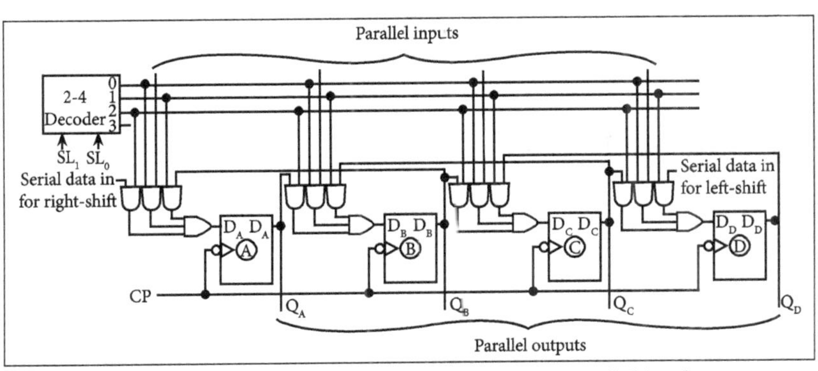

(b) 4-bit bi-directional shift register with parallel load.

5.2 Memories and ROM Architecture

Read-Only Memory (ROM) is the primary memory unit of any computer system along with the Random Access Memory (RAM), but unlike RAM, in ROM, the binary information is stored permanently. Now, this information to be stored is provided by the designer and is then stored inside the ROM. Once, it is stored, it remains within the unit, even when power is turned off and on again.

The information is embedded in the ROM, in the form of bits, by a process known as programming the ROM. Here, programming is used to refer to the hardware procedure which specifies the bits that are going to be inserted in the hardware configuration of the device. And this is what makes ROM a Programmable Logic Device (PLD).

ROM Organization

It is a read only memory. We can't write data in this memory. It is non-volatile memory i.e., it can hold data even if power is turned off.

An m x n ROM is an array of binary cells organized into m words of n bits each. As shown in the block diagram below, a ROM has k address input lines to select one of 2k = m words of memory, and n input lines, one for each bit of the word. An integrated circuit ROM may also have one or more enable inputs for expanding a number of packages into a ROM with larger capacity.

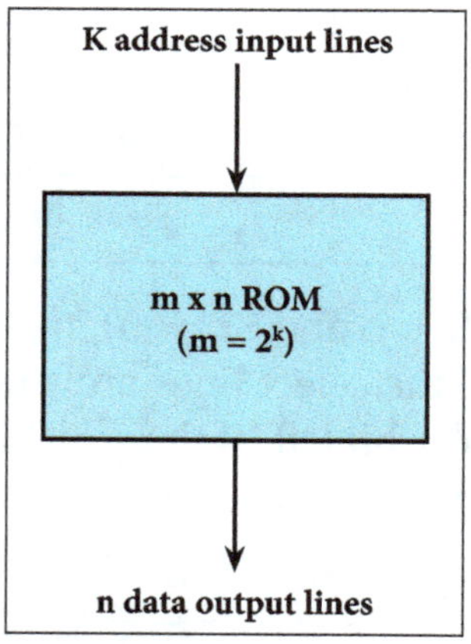

Read Only Memory.

ROM as Programmable Logic Device

A Programmable Logic Device (PLD) is an IC (Integrated Circuit) with internal logic gates connected through electronic paths that behave similar to fuses. In the original state, all the fuses are intact, but when we program these devices, we blow away certain fuses along the paths that must be removed to achieve a particular configuration. And this is what happens in ROM, ROM consists of nothing but basic logic gates arranged in such a way that they store the specified bits.

Typically, a PLD can have hundreds to millions of gates interconnected through hundreds to thousands of internal paths. In order to show the internal logic diagram of such a device a special semiology is used, as shown below:

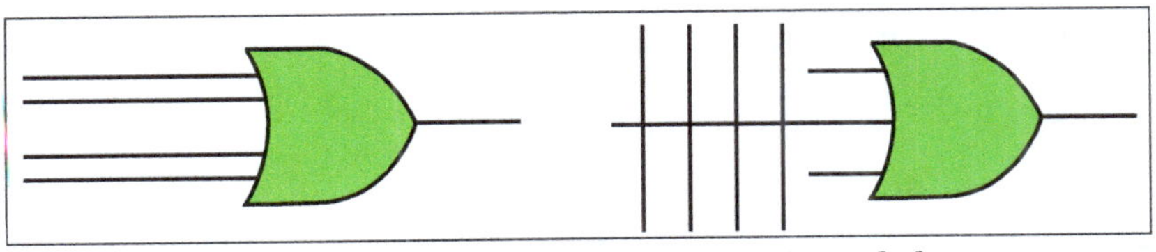

(a) Conventional Symbol (b) Array logic symbol.

The above figure shows the conventional way of representing inputs to a logic gate and the second symbol shows the special way of showing inputs to a logic gate called as Array Logic Symbol, where each vertical line represents the input to the logic gate.

Structure of ROM

The block diagram for the ROM is given below:

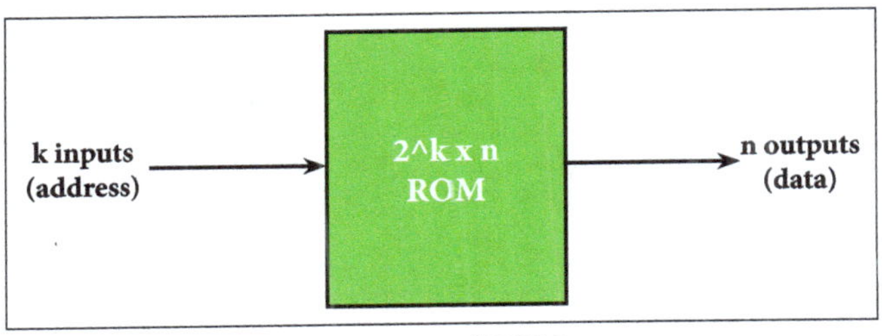

Structure of ROM.

Block Structure

- It consists of k input lines and n output lines.
- The k input lines is used to take the input address from where we want to access the content of the ROM.
- Since each of the k input lines can be either 0 or 1, so there are 2k total addresses which can be referred to by these input lines and each of these addresses contains n bit information, which is given out as the output of the ROM.
- Such a ROM is specified as 2^kxn ROM.

Internal Structure

- It consists of two basic components – Decoder and OR gates.
- A Decoder is a combinational circuit which is used to decode any encoded form (such as binary, BCD) to a more known form (such as decimal form).
- In ROM, the input to a decoder will be in binary form and the output will represent its decimal equivalent.

- The Decoder is represented as l x 2l, that is, it has l inputs and has 2l outputs, which implies that it will take l-bit binary number and decode it into one of the 2l decimal number.
- All the OR gates present in the ROM will have outputs of the decoder as their input.

5.3 Types of ROMS and its Application

There are four types of ROM such as:

- Masked ROM
- PROM
- EPROM
- EEPROM or E2PROM

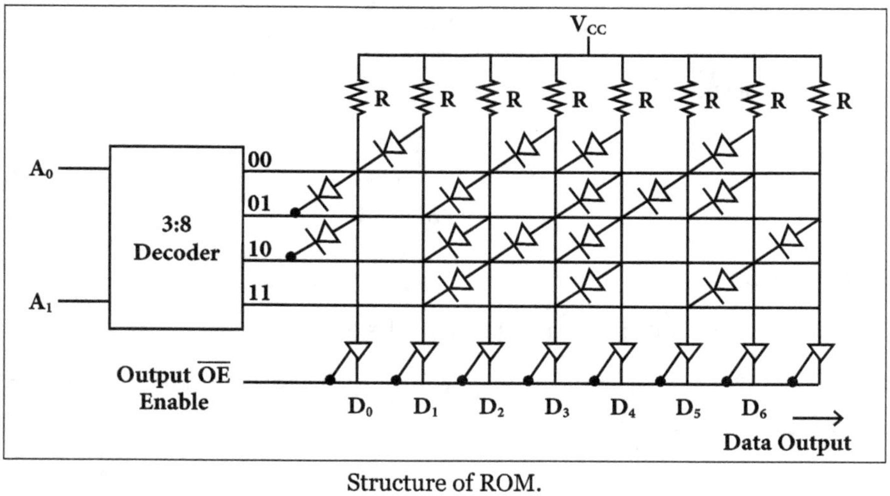

Structure of ROM.

Masked ROM

In masked ROM, mask programming is done by the manufacturer during the last fabrication process of the unit. The procedure for fabricating a ROM requires that the customer fill out the truth table which the customer wants the ROM to satisfy.

PROM

Bits can be set on a programming bench by burning fusible links or equivalent. This technology is also used for programmable array logic (PALs). The below figure shows four byte PROM. It has diodes in every bit position therefore, the output is initially all 0's.

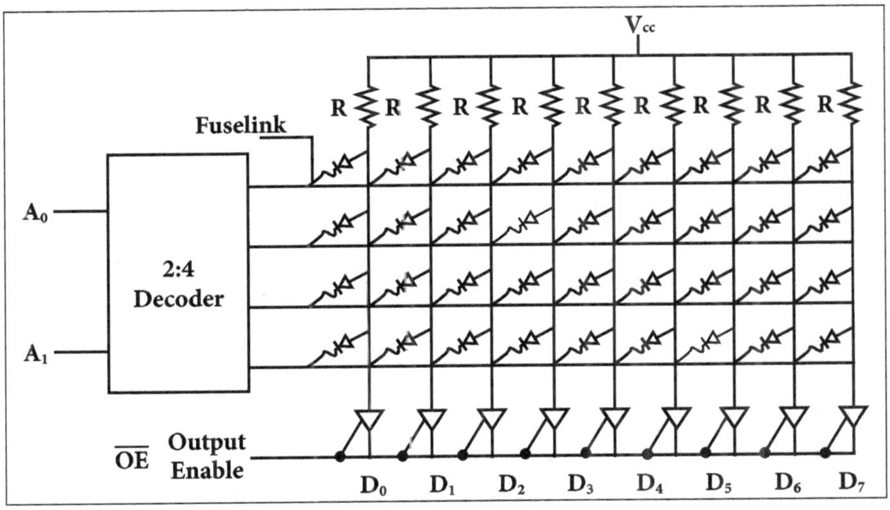

PROM

Each diode however has a fusible link in series with it. By addressing bit and applying proper current pulse at the corresponding output, we can blow out the fuse, storing logic 1 at that bit position. The PROM programmer selectively burns the fuses according to the bit pattern to be stored. This process is known as burning of PROM. It is one time programmable. Once programmed, the information stored is permanent.

EPROM: Erasable Programmable ROM

It is a type of ROM which can be erased using ultraviolet light. The erasable PROM (EPROM) is also useful for developmental and experimental work because its contents can be programmed with relative ease and at small cost.

The principal feature of the EPROM is that its contents can be erased and reprogrammed, enabling the same device to be used repeatedly. The process of programming an EPROM involves the application of special voltage levels to the appropriate chip inputs for a specified amount of time.

The complete programming process can take up several minutes for one EPROM chip. To erase an EPROM, the cells must be exposed to relatively intense UV radiation and it has transparent quartz window. This erasing process typically requires 15 to 30 minutes of exposure to UV rays.

Once erased, the EPROM can be reprogrammed. Once an EPROM has been programmed an opaque cable is placed over the window to prevent stray radiation from causing gradual erasure. EPROMs are available in a wide range of capacities and access times. A popular line of commercially available EPROMs is the 27xx series.

Difference between PROM and EPROM:

PROM	EPROM
PROM is Programmable Read Only Memory. It consists of a set of fixed AND gates connected to a decoder and a programmable OR array. The programmable OR gates implement the Boolean functions in sum of min-terms.	EPROM is Erasable and Programmable Read Only Memory. The ROM can be erased and programmed again and again.

EEPROM: Electronically Erasable Programmable ROM

ROM which can be erased electronically. It also use MOS circuitry similar to that of ROM. Data is stored as charge or no charge on an insulating layer, which is made very thin (< 200 A). Therefore, a voltage as low as 20 to 25 V can be used to move charges across the thin barrier in either direction for programming ROM or erasing.

It allows selective erasing at the register level rather than erasing all the information, since the information can be changed by using electrical signals. It has chip erase mode by which the entire chip can be erased in 10 ms. Hence, EEPROMs are most expensive and the least dense ROMs.

Electrically Alterable or Erasable PROM (EAPROM or EEPROM)

The EAPROM or EEPROM can be programmed (modified) or erased while it is still in the circuit and used like a nonvolatile read/write memory. EAPROM/EEPROM uses an electric charge to erase the contents.

Some types of EAPROM/EEPROM are more versatile. Individual cells can be reprogrammed by reversing the voltage used to create a zero. There are some timing constraints that causes more time for erasure or programming than is needed to read data from the part. Some EAPROM/EEPROM has a word or byte erase mode.

5.3.1 Applications

ROM is not volatile memory. It stores the data permanently.

The applications are as follows:

- ROMs are also used extensively in calculators and peripheral devices such as laser printers, which store their fonts.
- ROM is used in many appliances, toys, automobiles uses chips to maintain information when the power is shut off.
- ROM is used in embedded systems or any systems where the programming does not need to change.

5.4 RAM Architecture, Static and Dynamic RAMs

A memory unit is a set of storage cells together with associated circuits required to transfer information into and out of a device. The time it takes to transfer information to or from any desired random location is always the same. Hence, the name random access memory, abbreviated as RAM.

A memory unit keeps binary information in groups of bits called words. A memory word is a group of 1's and 0's and may represent a number, an instruction, one or more alphanumeric characters or any other binary-coded information. A group of 8 bits is called a byte. Most computer memories use words that are multiples of 8 bits in length.

Each word in memory is assigned an identification number called an address starting from 0 up to 2k -1, where k is the number of address lines. Consider for example, a memory unit with a capacity of 1K words of 16 bits each. Since 1K = 1024 = 210 and 16 bits constitute 2 bytes, we can say that the memory can accommodate 2048 = 2K bytes.

Memory address		
Binary	**Decimal**	**Memory content**
0000000000	0	1011010101011101
0000000001	1	1010101110001001
0000000010	2	0000110101000110
	⋮	⋮
1111111101	1021	1001110100010100
1111111110	1022	0000110100011110
1111111111	1021	1101111000100101

Contents of a 1024 × 16 memory.

In general, RAM refers to random access memory. The term RAM has also come to mean a memory which can be both easily written to and read from.

There are two main technologies used for RAM:

Static RAM

These are arrays of flip-flops. They can be fabricated in ICs as large arrays of thin flip-flops. "SRAM" is intrinsically somewhat faster than dynamic RAM.

Dynamic RAM

It uses capacitor arrays. Charge put on a capacitor will produce a HIGH bit if its voltage V = Q/C exceeds the threshold for the logic standard in use. Since the charge will "leak" o through the resistance of the connections in times of order 1msec, the stored information must be continuously refreshed (hence the term "dynamic").

Dynamic RAM can be fabricated with more bits per unit area in an IC than static RAM. Hence, it is usually the technology of choice for most large-scale IC memories.

Read and Write Operations

The two operations that RAM can perform are the write and read operations. The steps which are taken for storing a new word in the memory are as follows:

- Apply the binary address of the desired word to the address lines.
- Apply the data bit which is stored in memory as data input lines.
- Activate the write input.

The memory unit will take the bits from the input data lines and store them in the word specified by the address lines. The steps that should be taken for the purpose of transferring a stored word out of memory are as follows:

- Apply the binary address of the desired word to the address lines.
- Activate the read input.

Memory Enable	Read /Write	Memory Operation
0	X	None
1	0	Write to select word
1	1	Read from selected word

Control Inputs to Memory Chip

The memory enable (sometimes called the chip select) is used to enable the particular memory chip in a multichip implementation of a large memory.

When the memory enable is inactive, the memory chip is not selected and no operation is performed. When the memory enable input is active, the read/write input finds the operation to be performed.

Read Operation of SRAM

The complementary bit lines are pre charged to an intermediate voltage between VDD and ground. The bit lines are connected to the latches through access transistors when the cell is selected.

The differential voltage develops between complementary bit lines. The voltage changes and it should be sufficiently small not to change the latch state during readout. The read operation in an SRAM is nondestructive.

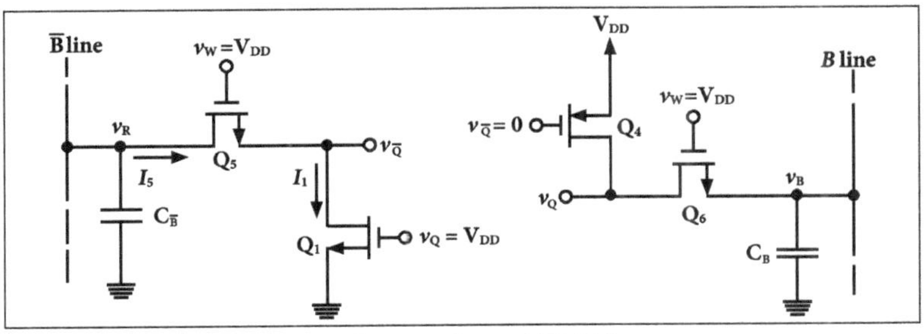

Read operation of SRAM.

Readout delay is determined by the rise time of word line and time needed to develop the required voltage. The dynamic operation can be approximated by treating Q_1 and Q_5 as r_{ON1} and r_{ON5}.

Design Constraint for Read Operation

The worst case scenario is to choose a pre charge value of VDD. Q5 operates in saturation region and Q1 operates in triode region:

$$I_5 = \frac{1}{2}\mu_n C_{ox}\left(\frac{W}{L}\right)_5 \left(V_{DD}-V_{in}-V_{\bar{Q}}\right)^2 = I_1 = \mu_n C_{ox}\left(\frac{W}{L}\right)_I \left[\left(V_{DD}-V_{in}\right)V_{\bar{Q}} - \frac{1}{2}V^2_{\bar{Q}}\right]$$

The design constraint is specified by,

$$V_{\bar{Q}} = \left(V_{DD}-V_{in}\right)\left(1-\frac{1}{\sqrt{1+\frac{(W/L)_5}{(W/L)_1}}}\right) \le V_{in} \rightarrow \frac{(W/L)_5}{(W/L)_1} \le \frac{1}{\left(1-\frac{V_{in}}{V_{DD}-V_{in}}\right)^2} - 1$$

The voltage change in vQ will be very small. The line is discharged by I5 at the beginning,

$$\Delta V = I_5 \Delta t / C_{\bar{B}}$$

The time needed to develop a voltage difference ΔV is given by,

$$\Delta t = C_{\bar{B}} \Delta V / I_5$$

The $\bar{B}$ line will be finally discharged to 0 in steady state. Sense amplifiers are usually used to speed up the read operation.

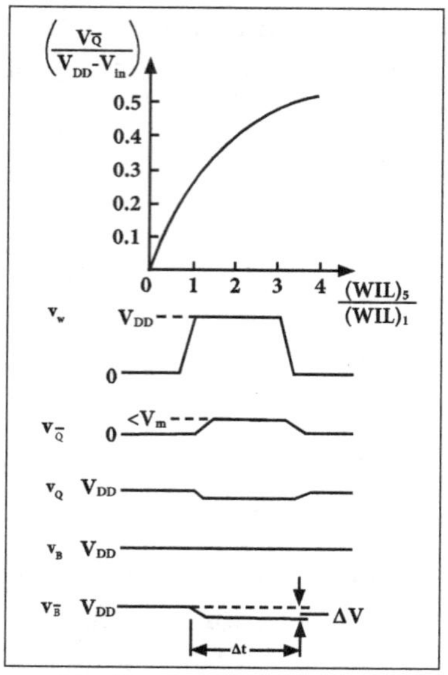

Read Operation of SRAM.

Write Operation of SRAM

The complementary bit-lines are respectively set to V_{DD} and ground before the operation. The cell is selected by the word line and the cell is connected to the bit lines through access transistors. Q_5 is in saturation and the current for charging capacitor equals to $I_5 - I_1$. I_5 decreases (due to reduced VGS_5 and body effect) and I1 increases (due to reduced VGS_1). When v_Q and reach $V_{DD}/2$, the regenerative feedback initiates and cause the flip-flop to change state.

Write delay is determined by the time for v_Q and $v_{\bar{Q}}$ to reach $V_{DD}/2$ and the delay time of the flip-flop. The charging (discharging) component of write delay is much smaller than the corresponding component in the read operation because only small capacitance C_Q needs to be charged (discharged). The delay time in write operation is dominated by the word-line delay.

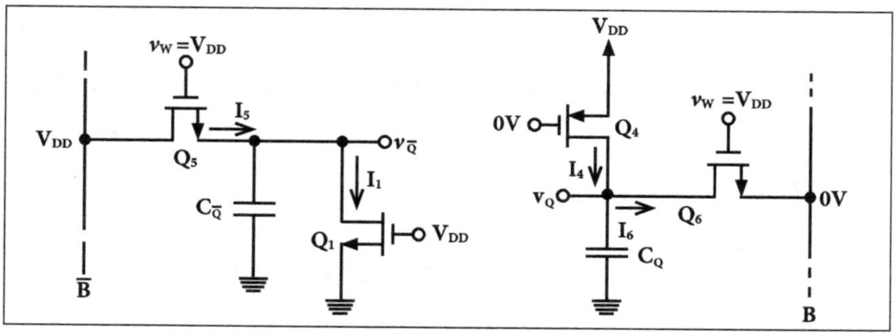

Write operation of SRAM.

Design Constraint for Write Operation

Q_4 operates in saturation region,

$$I_5 = \frac{1}{2}\mu_n C_{ox}\left(\frac{W}{L}\right)_4 \left(V_{DD} - |V_{in}|\right)^2 = I_6 = \mu_n C_{ox}\left(\frac{W}{L}\right)_I \left[\left(V_{DD} - V_{in}\right)V_Q - \frac{1}{2}V_Q^2\right]$$

The design constraint is specified by,

$$V_Q = \left(V_{DD} - V_{in}\right)\left(1 - \sqrt{\frac{\mu_p (W/L)_4}{\mu_n (W/L)_6}}\right) \le V_{in} \rightarrow \frac{(W/L)_4}{(W/L)_6} \le \frac{\mu_p}{\mu_n}\left[1 - \left(1 - \frac{V_{in}}{V_{DD} - V_{in}}\right)^2\right]$$

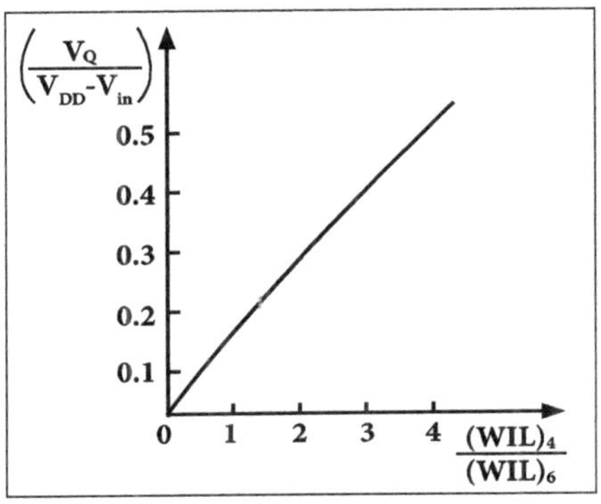

Characteristic Curve.

Timing Diagram

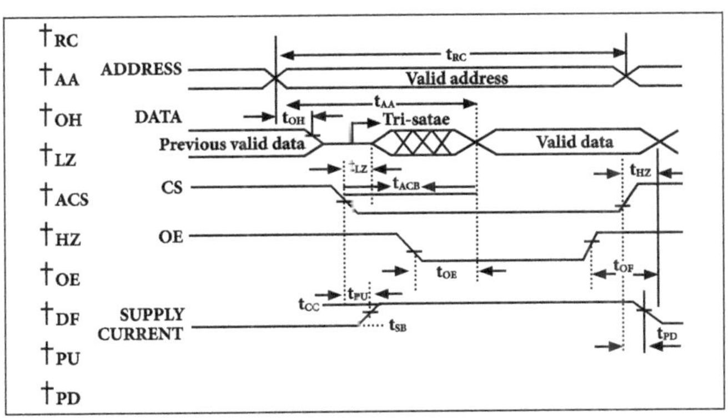

Read cycle.

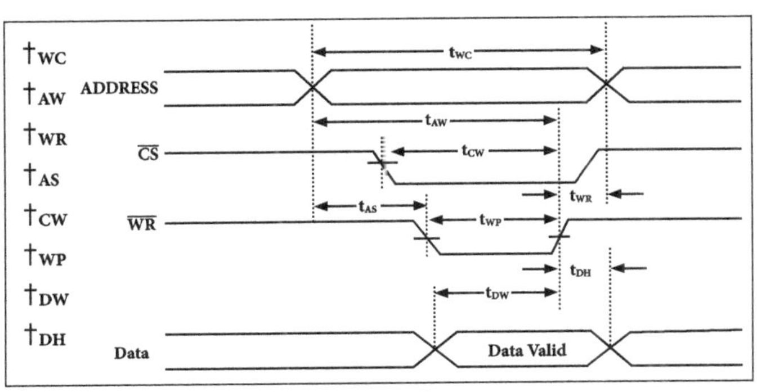

Write cycle.

Dynamic RAM Cell

Read Operation: The bit lines are pre charged to VDD/2. The word line is raised to VDD and all the access transistors in the selected row are conductive. A small voltage difference will be detected in the bit lines. The read operation is destructive.

$$C_S V_{CS} + C_B \frac{V_{DD}}{2} = (C_B + C_S)\left(\frac{V_{DD}}{2} + \Delta V\right)$$

$$\Delta V = \frac{C_S}{C_B + C_S}\left(V_{CS} - \frac{V_{DD}}{2}\right) \approx \frac{C_S}{C_B}\left(V_{CS} - \frac{V_{DD}}{2}\right) \text{ for } C_B >> C_S$$

Write Operation: The data bit to be written is set to V_{DD} or ground and CS will be charged or discharged to V_{DD} -V or 0V. All other cells in the selected row perform write operation at the same time.

Refresh Operation: Read operation followed by a write operation for stored data.

Memory Decoding

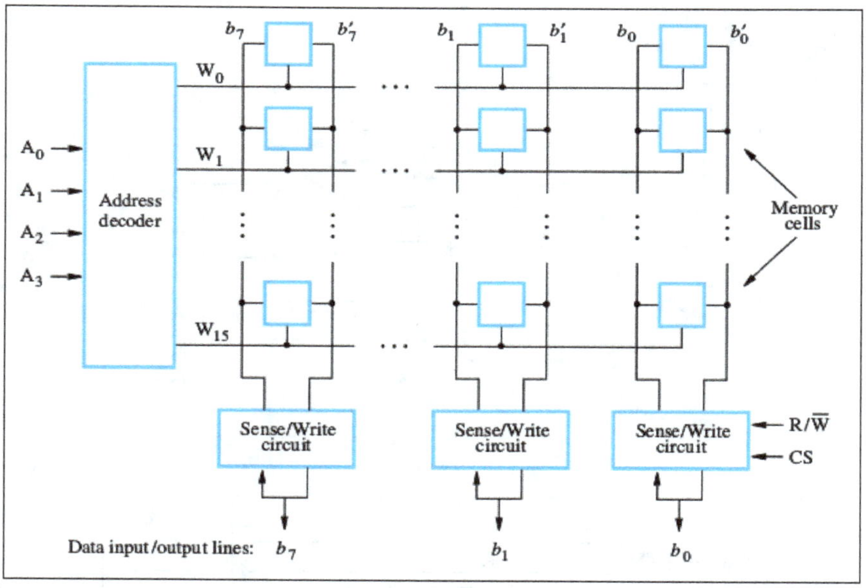

Memory Decoding.

Decoder Operation:

- It has 16 words of 8 bits each.
- A memory with 16 words needs 4 address lines.
- Here, 4 x 16 decoder is used.

- If memory enable = 0, no memory word is selected.
- If memory enable = 1, one of the 16 words is selected.
- Read/write command determines the operation.

Write Operation:

- Data is transferred into eight memory cells.
- If a memory cell is not selected, the previous value remains unchanged.

Coincident Decoding

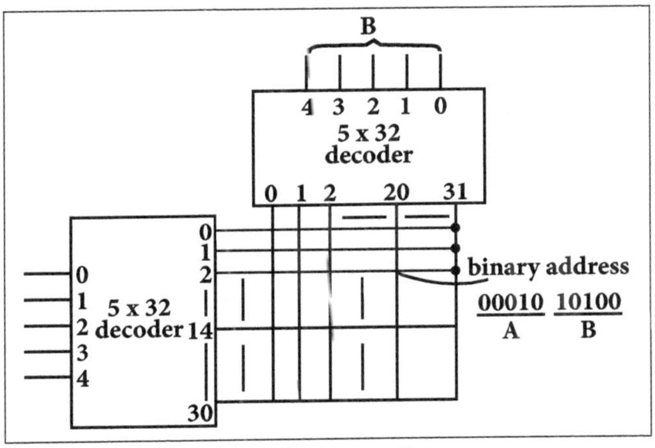

Coincident Decoding.

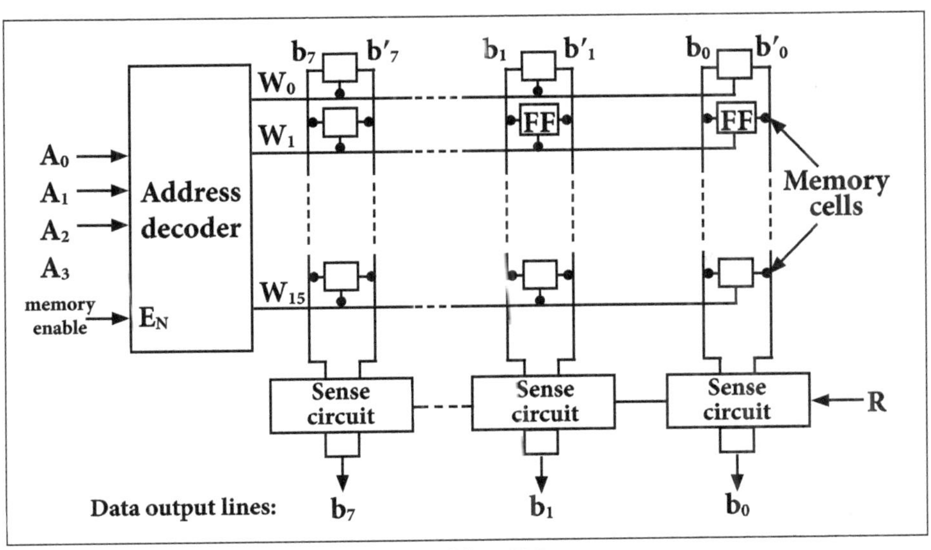

16 x 8 bit ROM.

- It has k input and 2k output.
- 2k input gates are needed with a k input to each gate.

- 2 decoders are used to reduce the number of inputs.
- K/2 inputs are given to each decoder instead of k inputs.
- Instead of 10 x 1024 decoder, we use 5 x 32 decoders.

Error Detection and Correction

Error Detecting Codes

When the digital information in the binary form is transmitted from one circuit or system to another circuit or system, an error will occur. To maintain the data integrity between the transmitter and receiver, the extra bit or more than one bit is added to the data. The data with the extra bits forms the code. Codes which allow only error detection is called as error detecting codes.

A signal corresponding to 0 changes to 1 or vice-verse due to the presence of noise. To maintain the data integrity between the transmitter and receiver, an extra bit or more than one bit is added in the data.

These extra bits allow the detection and sometimes correction of error takes place in the data. The data with the extra bits forms the code. Codes which allow only error detection is called as error detecting codes and codes which allow both error detection and error correction are called as error detecting and error correcting codes.

The most common cause for errors is the noise creep in the bit stream during the course of transmission from transmitter to the receiver.

If these errors are not detected and corrected, the result will be worst where the digital systems are too much sensitive to errors and will malfunction due to the errors in the transmitted codes.

The various methods of error detection and correction such as addition of extra bits is called as check bits which is also called as redundant bits as they don't have any information in them.

The various codes used for error detection and correction in the digital system are as follows:

- Two-dimensional parity check.
- Simple parity check.
- Checksum.
- Cyclic redundancy check.

Error Correcting Codes

Error correction can be handled in two ways.

They are:

- When an error is discovered, the receiver makes the sender to retransmit the entire data unit. This is termed as backward error correction.
- The receiver can use an error-correcting code which automatically corrects certain errors. This is known as forward error correction.

It is possible to correct any number of errors automatically. The error correcting codes are more sophisticated than the error detecting codes and they need more redundant bits. The number of bits required to correct the multiple bits or burst error is so high. For this reason, most error correction is limited to one, two or three bits.

Single-Bit Error Correction

The concept of error-correction is understood by examining the simplest case of single-bit errors. A single-bit error is detected by addition of a parity bit (VRC) with the data.

A single additional bit can detect an error, but it is not enough to correct that error. For collecting an error, one has to know the exact position of error i.e. exactly which bit is in the error.

For example, to correct the single-bit error in an ASCII character, the error correction should determine which one of the seven bits is in the error. To do this, we have to add some additional redundant bits.

For calculating the number of redundant bits (r) required to correct (d) data bits, let us determine the relationship between the two. Therefore, we have (d+r) as the total number of bits which are to be transmitted where r indicates at least d+r+1 value.

One value means no error and remaining d+r values indicate error location of the error in each of the d+r locations. So, d+r+1 state must be distinguishable by r bits and r bits indicates 2r states. Thus, 2r must be greater than d+r+1, i.e. $2r > d + r + 1$.

The value of r is determined by putting the value of d in the above relation.

Example: If d is 7, then the smallest value of r that satisfies the above relation is 4. Thus, the total bits which are to be transmitted is 11 bits ($d + r = 7 + 4 = 11$).

Parity Checking

Parity Code

It is the simplest technique which is used for detecting and correcting errors. The MSB of an 8-bits word is used as the parity bit and the remaining 7 bits are used for data or message bits. Parity bit is added to the transmitted strings of bits during the transmission from the transmitters to detect any error in the data.

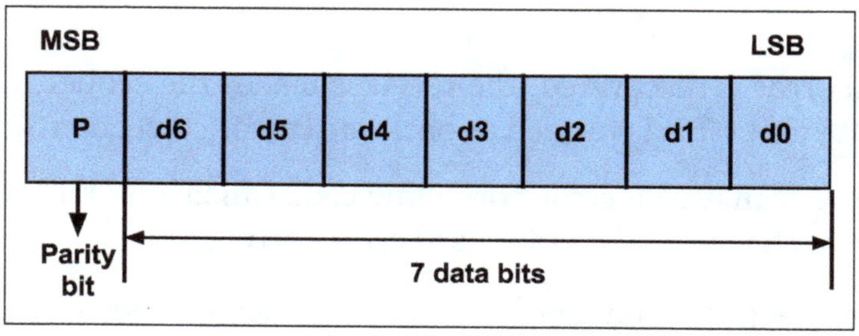

Parity Checking.

When they are received at the receiver end, an extra bit is added to the string of data. The parity of 8-bits transmitted word can be either even parity or odd parity.

Parity checking at the receiver detects the presence of an error if the parity of receiver signal is different from the expected parity. If it is known that the parity of the transmitted signal is always even and if the received signal has an odd parity, then the receiver can conclude that the received signal is not correct. When an error is detected, then the receiver will ignore the received byte and request for the retransmission of the same byte to the transmitter.

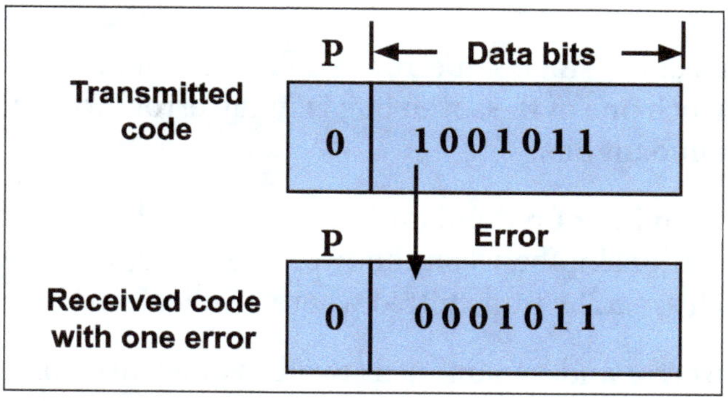

Parity Code.

Even Parity

Even parity means the number of 1's in the given word including the parity bit should be even. When we get an even parity after adding that extra bit, the total numbers of 1 in the string of the data is even.

Odd Parity

Odd parity is defined as the number of 1's in the given word including the parity bit which is odd. When we get an odd parity after adding that extra bit into the data string, the total number of 1's in the data is odd.

Let us consider an example where we have an eight bit ASCII code as 01000001.

If the added bit is 0, then the number will become 001000001. In odd parity, if the total number of 1's in the number is even, and then we get an even parity. If we add 1 to the number, then the number will become 101000001.

The number of 1's is 3 is odd which results in an odd parity. The parity checks are capable of detecting the single bit error. If there are two changes in the data, it is the biggest drawback of this system. Hence, there are several other codes to detect and correct more than one bit errors.

Use of a Parity Bit

The parity bit can be set to 0 and 1 which depends on the type of the parity required.

For even parity, this bit is set to 1 or 0 where the number of 1 bits in the entire word is even which is shown in the below figure (a).

For odd parity, this bit is set to 1 or 0 whereas the number of 1 bits in the entire word is odd which is shown in the below figure (b).

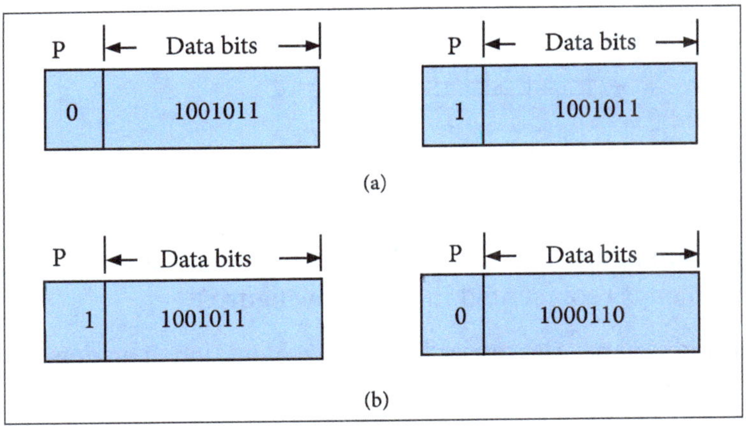

Parity bit.

Hamming Code

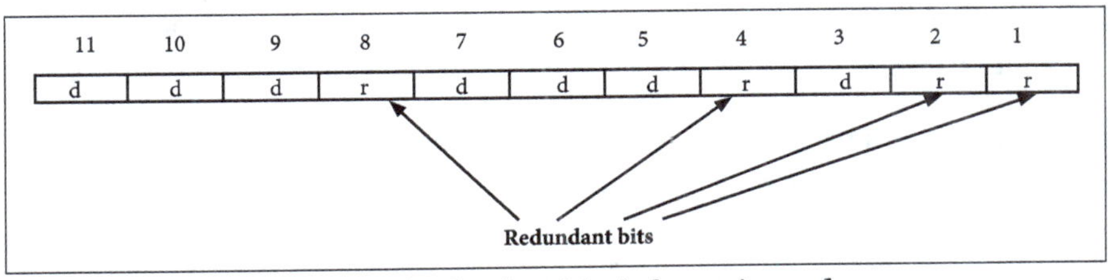

Positions of redundant bits in hamming code.

Let us examine the manipulation with these bits to discover which bit is in the error. The coding scheme developed by R.W. Hamming is called as hamming code. The hamming code is applied to data units of any length and it uses the relationship between the data bits and the redundant bits.

The basic method for error detection by using hamming code is as follows:

- The location of each of the m + K digits is assigned a decimal value.

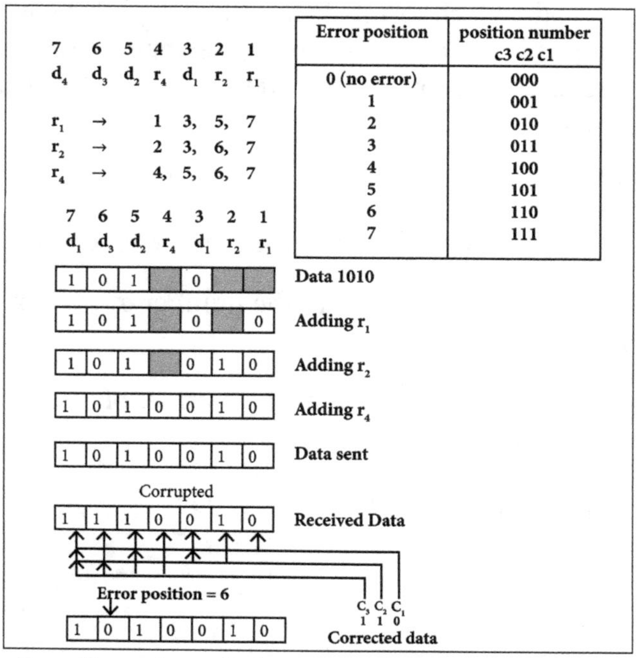

Use of hamming code for error collection for a 4-bit data.

- To each group of 'm' information bits, 'k' parity bits are added to form (m + k) bit code which is represented in the below figure.
- At the receiving end, the parity bits are recalculated. The decimal value of the k parity bits indicates the bit-position in error.

The above figure represents how the hamming code is used for the correction of 4-bit numbers ($d_4d_3d_2d_1$) with the help of three redundant bits (r3r2r1). For example, consider data 1010. First, r1 is determined by considering the parity of the bit positions 1, 3, 5 and 7. Then, the parity bits r_2 is calculated by considering bit positions 2, 3, 6 and 7. Finally, the parity bits r4 is calculated by considering the bit positions 4, 5, 6 and 7.

If any corruption occurs in any of the transmitted code 1010010, the bit position in the error is found by calculating r3r2r1 at the receiving end. For example, if the received code word is 1110010, the recalculated value of r3r2r1 is 110 which indicate that bit position in error is 6, and then the decimal value is 110.

- Parity generator: It is the circuit which generates the parity bit in the transmitter.
- Parity checker: It is the circuit that checks the parity in the receiver.

5.4.1 Static and Dynamic RAMs

Types of RAM

Integrated circuit RAM units are available in 2 operating modes.

They are as follows:

- Static mode.
- Dynamic mode.

Static RAM (SRAM)

It has internal latches which stores the binary information. The stored information remains valid as long as power is applied to the unit.

Dynamic RAM (DRAM)

It stores the binary information in the form of electric charges on capacitors inside the chip by MOS transistors. The stored charge on the capacitors tends to discharge with time and the capacitors are periodically recharged by refreshing the dynamic memory.

Refreshing is done by cycling the words for every few milliseconds and restores the decaying charge. DRAM gives reduced power consumption and larger storage capacity in a single memory chip. SRAM is simple to use and has shorter read and write cycles.

ROM is another nonvolatile memory. A nonvolatile memory allows the digital computers to store programs that will be required again after the computer is turned on. Programs and data which can't be altered are stored in ROM while other large programs are maintained on magnetic disk.

Permissions

All chapters in this book are published with permission under the Creative Commons Attribution Share Alike License or equivalent. Every chapter published in this book has been scrutinized by our experts. Their significance has been extensively debated. The topics covered herein carry significant information for a comprehensive understanding. They may even be implemented as practical applications or may be referred to as a beginning point for further studies.

We would like to thank the editorial team for lending their expertise to make the book truly unique. They have played a crucial role in the development of this book. Without their invaluable contributions this book wouldn't have been possible. They have made vital efforts to compile up to date information on the varied aspects of this subject to make this book a valuable addition to the collection of many professionals and students.

This book was conceptualized with the vision of imparting up-to-date and integrated information in this field. To ensure the same, a matchless editorial board was set up. Every individual on the board went through rigorous rounds of assessment to prove their worth. After which they invested a large part of their time researching and compiling the most relevant data for our readers.

The editorial board has been involved in producing this book since its inception. They have spent rigorous hours researching and exploring the diverse topics which have resulted in the successful publishing of this book. They have passed on their knowledge of decades through this book. To expedite this challenging task, the publisher supported the team at every step. A small team of assistant editors was also appointed to further simplify the editing procedure and attain best results for the readers.

Apart from the editorial board, the designing team has also invested a significant amount of their time in understanding the subject and creating the most relevant covers. They scrutinized every image to scout for the most suitable representation of the subject and create an appropriate cover for the book.

The publishing team has been an ardent support to the editorial, designing and production team. Their endless efforts to recruit the best for this project, has resulted in the accomplishment of this book. They are a veteran in the field of academics and their pool of knowledge is as vast as their experience in printing. Their expertise and guidance has proved useful at every step. Their uncompromising quality standards have made this book an exceptional effort. Their encouragement from time to time has been an inspiration for everyone.

The publisher and the editorial board hope that this book will prove to be a valuable piece of knowledge for students, practitioners and scholars across the globe.

Index

www.ingramcontent.com/pod-product-compliance
Lightning Source LLC
LaVergne TN
LVHW080607210926
848485LV00003B/112